Third Edition

Making Sense

A Student's Guide to
Research and Writing

Life Sciences

Margot Northey

Patrick von Aderkas

OXFORD
UNIVERSITY PRESS

Oxford University Press is a department of the University of Oxford.
It furthers the University's objective of excellence in research, scholarship,
and education by publishing worldwide. Oxford is a registered trade mark of
Oxford University Press in the UK and in certain other countries.

Published in Canada by
Oxford University Press
8 Sampson Mews, Suite 204,
Don Mills, Ontario M3C 0H5 Canada

www.oupcanada.com

Copyright © Oxford University Press Canada 2019

The moral rights of the authors have been asserted
Database right Oxford University Press (maker)

First Edition published in 2011
Second Edition published in 2015

All rights reserved. No part of this publication may be reproduced, stored in
a retrieval system, or transmitted, in any form or by any means, without the
prior permission in writing of Oxford University Press, or as expressly permitted
by law, by licence, or under terms agreed with the appropriate reprographics
rights organization. Enquiries concerning reproduction outside the scope of the
above should be sent to the Permissions Department at the address above
or through the following url: www.oupcanada.com/permission/permission_request.php

Every effort has been made to determine and contact copyright holders.
In the case of any omissions, the publisher will be pleased to make
suitable acknowledgement in future editions.

Library and Archives Canada Cataloguing in Publication

Northey, Margot, 1940–, author
    Making sense in the life sciences : a student's guide to writing and
research / Margot Northey, Patrick von Aderkas.—Third edition.

(Making sense)
First edition published under title: Making sense : life sciences : a
student's guide to research and writing.
Includes bibliographical references and index.
Issued in print and electronic formats.
ISBN 978-0-19-902681-4 (softcover).—ISBN 978-0-19-902674-6 (epub)

1. Life sciences—Authorship. 2. Life sciences—Research. 3. Report
writing. 4. English language—Rhetoric. I. Aderkas, P. von, author
    II. Title. III. Series: Making sense series

QH304.N67 2018   808.06'657   C2018-901502-0
                                            C2018-901503-9

Oxford University Press is committed to our environment.
This book is printed on Forest Stewardship Council® certified paper
and comes from responsible sources.

Cover image: Andrew Brooke/Cultura/Getty Images
Cover design: Laurie McGregor
Interior design: Sherill Chapman

Printed and bound in the United States of America

1 2 3 4 — 21 20 19 18

Contents

Acknowledgements   iv
A Note to the Student   v
A Note to the Instructor   vi

1. Writing and Thinking   1
2. Planning an Essay   12
3. Researching an Essay   26
4. Writing an Essay   50
5. Writing a Lab Report   89
6. Writing with Style   106
7. Common Errors in Grammar and Usage   125
8. Punctuation   140
9. Misused Words and Phrases   156
10. Using Illustrations   171
11. Documenting Sources   188
12. Giving Oral Presentations and Poster Presentations   203
13. Working in Groups   232
14. Writing Examinations   243
15. Writing Resumés and Letters of Application   256

Appendix: Weights, Measures, and Notation   267
Glossary   268
Index   274

Acknowledgements

I am really grateful for suggestions from a number of undergraduate students (Gillian Cook, Caitie Frenkel, Stephanie Korolyk, Barry Macdonald, Dani Sweetnam-Holmes, Marie Vance, Catherine Weiss, and Brett Yerex) and graduate students (Andrea Coulter and Natalie Prior). One of the nice surprises was the detailed commentary provided by Frédérique Guinel of Wilfrid Laurier University (WLU).

It has been a pleasure working with the editorial staff of OUP. Marta Kule and Lauren Wing have been wonderful developmental editors. I am particularly grateful to Jessie Coffey whose meticulous copy-editing lifted this book. I would also like to thank reviewers Wilma Groenen, Afeez Hazzan, and Julia Wallace, as well as the anonymous reviewers, whose clever suggestions I incorporated into this incarnation of *Making Sense*.

I dedicate this edition to my wife, Elizabeth, who kept me in writerly good cheer.

Patrick von Aderkas
University of Victoria

A Note to the Student

Life sciences students live crazy busy lives. There are exams, lectures, and labs. As deadlines pile up, topics have to be researched, digested, and then thoughtfully written up. A popular belief is that it takes 10,000 hours to master any skill. But you don't have 10,000 hours! This text presents you with concise information that will substantially reduce the time it takes you to acquire mastery in writing. We present practical advice to help you overcome common pitfalls in grammar, style, punctuation, and usage. We provide a rich variety of examples that will help you improve your writing. We even include an annotated problem essay along with its revised version to show you how you can improve your own essays. We have also thoroughly updated the writing examples to be even more relevant to the life sciences. You will find that once you gain a little bit of control, more follows, and eventually you will be able to write about science with fluidity and ease.

In this edition, we provide updated examples on how to find the papers that you need for your essays and how to accurately reference this information. We provide examples of how to search across various databases to find the important papers in a field, as well as how to burrow deeply into scientific literature.

Life sciences foster cooperative effort in labs and classes; in a scientific career, such cooperation leads to inter-lab collaborations and large-scale projects. To help you learn to work in groups, we have provided advice on lab partnerships, study groups, and other types of cooperative work.

A Note to the Instructor

A strength of the *Making Sense* series is its focus on essays, reports, presentations, and resumés. There are chapters to help students improve punctuation, grammar, documentation, word usage, and style. In this edition, we aimed at adapting this content to students in the life sciences. To this end, we added and enhanced material on lab reports, exam preparation, group work, graphics use, and literature searches.

We also took a fresh approach to research techniques. Students use search engines and browsers in unexpected and surprising ways. We're not alone in coming to the conclusion that methods of research have changed. Electronic databases not only allow student scholarship to develop along traditional lines—that is, to search ever deeper into a subject—but they also allow horizontal searches through citations, cross-references, and similar titles. This type of skimming forces students to alter their search strategies on the fly, which can lead to great success. In short, there is more than one good way to find information these days.

This book is much more than a writing guide. Many courses in the life sciences have little or no assigned written work and may even lack a lab component. Teaching in the life sciences involves getting students to talk about what they have learned. Often, the best way to get students to discuss what they know is to assign them group work. Consequently, this book contains useful advice to help students work collaboratively during classes and labs to complete assignments as well as in study groups to prepare for exams.

1 Writing and Thinking

Objectives
- Developing strategies for tackling a writing project
- Defining your purpose
- Avoiding problems in style and tone
- Thinking like a scientist—the value of questions

You are not likely to produce clear writing unless you have first done some clear thinking, and thinking can't be hurried. It follows that the most important step you can take is to leave yourself enough time to think. Psychologists have shown that you can't always solve a difficult problem by "putting your mind to it"—by determined reasoning. Sometimes when you are stuck it's best to take a break, sleep on it, and let the subconscious or creative part of your brain take over for a while. Very often a period of relaxation will produce a new approach or solution. Just remember that leaving time for creative reflection isn't the same thing as sitting around waiting for inspiration to strike out of the blue.

Initial Strategies

Writing is about making choices: choices about what ideas you want to present and how you want to present them. Practice makes decision-making easier, but no matter how fluent you become, with each piece of writing you will still have to choose.

You can narrow the field of choice from the start if you realize that you are not writing for just anybody, anywhere, for no particular reason. With any writing you do, it's always a sound strategy to ask yourself two basic questions:

- What is the purpose of this piece of writing?
- What is the reader like?

Think about the purpose

Depending on the assignment, you may have any one (or more) of the following purposes for writing an essay:

- to show your knowledge of a topic or text;
- to show that you understand certain terms or theories;
- to show that you can do independent research;
- to show that you can apply a specific theory to new material;
- to demonstrate your ability to evaluate secondary sources; and/or
- to show that you can think critically or creatively.

An assignment designed to see if you have read and understood specific material requires a different approach from one that's meant to test your critical thinking. In the first case, your approach will tend to be *expository*, with the emphasis on presenting facts. In the second case, you will probably want to structure your essay around a particular argument or assertion that other people might dispute. Your aim in this kind of *argumentative* or *persuasive* essay is to bring your reader around to your point of view. (Argumentative essays are discussed in greater detail in Chapter 2.)

Think about the reader

To convince a reader that your own views are sound, you have to show that you not only understand the subject but can also formulate an articulate argument. You will have to make specific decisions about the terms you should explain, the background information you should supply, and the details you will need to provide. It pays to imagine two kinds of readers: the first one is interested only in the facts, whereas the second one expects you to build on the facts to produce a well-crafted, scholarly piece of writing. In university or college, satisfying the first type of reader may get you a pass, but trying to satisfy the second type will turn you into a very good writer. If you don't know who will be reading your paper—your professor, your tutorial leader, or a marker—just imagine someone intelligent, knowledgeable, and interested, who is skeptical enough to question your ideas but flexible enough to adopt them if your evidence is convincing.

Think about the audience

Not all writing that you do is for essays and lab reports. Some will be for presentations. To convince listeners that you know what you are talking about, you need to think about the audience. For most life science students that audience is characteristically one of two types. The first type is composed of fellow students who know very little or even nothing at all about the subject you are about to present. The other audience is fellow science majors. Your presentation will be based on materials that you have gathered and assembled to make your argument. Unlike essay writing, in which you assume you have only to please one reader (e.g., the marker), in the case of a presentation you have many listeners. These can include people who are marking your presentation, as well as engaged fellow students, disengaged fellow students, and possibly even personal friends. For these reasons, the pressures during a presentation are psychologically great. Presentations are the hardest things to write, because loss of clarity causes audience discomfort. Public speaking is a phobia for many students and requires experience if it is to be mastered. Although all writing must strive for clarity, writing for presentations requires great care, because you want to hold everyone's attention without publicly humiliating yourself.

An often overlooked aspect of writing for an audience is that listeners are not like readers. This is because listeners are in a social setting. The venue is not of their choosing. They, too, may be nervous about their own imminent presentations. In addition, distractions commonly occur during a presentation (e.g., a loud airplane flies overhead drowning you out, or a dropped object causes a distraction). In short, if you want an important point to stick in the listener's mind, you probably need to be prepared to make it in more than one way. You may even need to repeat your point verbatim. Repetition is, of course, a rhetorical device. The point is that writing for presentations allows you to draw on rhetorical devices that would be inappropriate for an essay.

Let's look at the ways that you need to think about your audience that will directly affect your writing of the presentation. Your opening should lay out why the subject should interest the listener. Having a vivid introduction is quite important. Take the case of a presentation to an audience that knows nothing about your assigned subject. These could be students in a first-year biology course that is a required course for quite different programs. Any given student may not be a life sciences major. In many ways, this is no different than being an interpreter who gives talks on natural history to park visitors. The assumptions of a good interpreter are that the visitors know nothing, that they are there to listen, and that they're willing to be engaged. The key to success is to win and to keep your listeners' attention. What works in your writing is sometimes self-evident. Fortunately, it is also very easy to test for effective presentation writing. Read what you have written out loud to yourself. If you don't like that option, read it into a recording device and play it back to yourself. Another option is to read it aloud to a friend. In all cases, it must be read in the same manner and pacing as if you were presenting to an audience. Flaws can be quickly detected: passages that are unclear or that show strains in the logic are immediately obvious. Boring bits quickly come to light. You can also hear whether your points are clearly made and where emphasis needs to be placed.

The second type of audience is one made up of fellow majors. Although all of the aforementioned suggestions apply to this audience, there is also the challenge of presenting more complex subject matter. The students are likely to be a more sympathetic audience, because they already understand many aspects of your subject. Interest in your subject is not as hard to win. However, you will likely be required to describe far more complex subject matter. For example, you may have to describe messy studies that contradict one another. It is possible to keep the listener interested by inserting, for example, rhetorical questions, or even pausing for effect. Effective delivery still requires the greatest clarity. Again, reading the written presentation aloud to yourself is a key component in improving your delivery.

Think about the length

Before you start writing, you will also need to think about the length of your assignment in relation to the time you can spend on it. If both the topic and the length are prescribed, it should be fairly easy for you to assess the level of detail required and the amount of research you will need to do. If only the length is prescribed, that restriction will help you decide how broad or how narrow a topic you should choose (see pages 19–22).

Think about the tone

In everyday writing to friends you probably adopt a casual tone, but academic writing is usually more formal. Just how formal you need to be will depend on the kind of assignment and on the instructions you have been given. Life sciences essays and reports require a formal tone.

What kind of style is too informal for most academic work? Here are the main signs.

Use of slang

Although the occasional slang word or phrase may be useful for special effect in a presentation, frequent use of slang is not acceptable in academic writing because slang expressions are usually regional and short-lived. They may mean different things to different people at different times. (Just think of how widely the meanings of *hot* and *cool* can vary, depending on the circumstances.) In a formal essay, where clarity of expression is important, it's better to use words with well-established meanings that will be understood by the greatest number of readers. If you think a word you're using might be slang, look it up in a dictionary or conduct an online search (try typing your word plus *slang* into your search engine to get the most relevant results). A quick way to find alternative words that are more formal and acceptable is to use a thesaurus (see page 107).

Excessive use of first-person pronouns

Because a formal essay is not a personal outpouring, you want to keep it from becoming *I*-centred. There is no need to begin every sentence with "I think" or "In my view" when the facts or arguments speak for themselves. Over the past 30 years, first-person pronouns have become more common in life sciences writing. First-person narration has even become acceptable in international peer-reviewed journals. This trend can confuse a student who has been told that science is best written about from a detached perspective. If you aren't sure what your instructor expects, it's always best to ask. It's certainly acceptable to use the occasional first-person pronoun if the assignment calls for your point of view—as long as your opinions are backed by evidence. You should never use *we* to include the reader—for example, "Here we see that . . ."

The scientific tone is achieved by taking the researcher (you) out of the research being reported. This often means using the passive voice:

> All fish were acclimated to laboratory conditions for a minimum of three weeks.

> Historically, species have been viewed as relatively fixed entities that cannot evolve in response to changing climate.

In both sentences, the writer is neither stating a personal opinion nor announcing his/her presence in the experiment. The writer is trying to create an objective tone. In short, he or she is objectively describing an experiment on fish, just as he or she is objectively describing his or her view of species in a dispassionate manner.

Also, if the choice is between using *I* and creating a tangle of passive constructions, it's almost always better to choose *I*. (A hint: when you do use *I*, it will be less noticeable if you place it in the middle of the sentence rather than at the beginning—this strategy is especially useful when writing resumés.) Here are some examples of ways to avoid *I*-centred and unnecessarily passive sentences:

- ✗ Having analyzed the fossils, I believe they are from two species.

- ✗ The fossils, having been analyzed, appear to me to be from two species.

- ✔ Analysis revealed that the fossils are likely from two species.

- ✔ [or] When analyzed, the fossils seem to be from two species.

- ✗ In this essay, Aldo Leopold's portrayal of nature will be investigated, and the repeated conflict between wilderness and civilization will be discussed.

- ✔ [better] In this essay, I will investigate Aldo Leopold's portrayal of nature and discuss the repeated conflict between wilderness and civilization.

- ✔ [best] This essay will investigate Aldo Leopold's portrayal of nature and discuss the repeated conflict between nature and civilization.

Frequent use of contractions

Generally speaking, contractions such as *can't* and *isn't* are not suitable for academic writing, although they may be fine for letters or other informal kinds of writing—for example, this handbook. This is not to say that you should avoid using contractions altogether; even the most serious academic writing can sound stilted or unnatural without any contractions at all. Just be sure that when you use contractions in a college or university essay you use them sparingly because excessive use of contractions makes formal writing sound chatty and informal.

Finding a suitable tone for academic writing can be a challenge. The problem with trying to avoid excessive informality is that you may be tempted to go to the other extreme. If your writing sounds stiff or pompous, you may be using too many inflated phrases, long words, or passive constructions (see Chapter 6). When in doubt, remember that a more formal style is the best option.

Think about the structure

Embarking on a career in the life sciences requires you to develop a diverse set of skills—you must master scientific techniques and methods, develop your analytical thinking, learn to design experiments, and ultimately apply these skills in order to discover aspects of nature. Underlying most of this process is an attitude that is based on a questioning spirit. The sooner you learn to approach essays and assignments in this spirit, the faster you will develop a recognizably scientific writing style. The trick is to think like a scientist. Simply put, you must question everything.

In your first year of study, you are given *descriptive* assignments, but very soon you will be given more *analytical* assignments. A descriptive assignment is one in which you describe something you have done, or you have to record observations on a process or an organism. An analytical assignment involves experimentation. You may be asked to test a hypothesis, to question the truth of a statement. A solid strategy for approaching analytical assignments involves asking questions of increasing sophistication. The first questions that you should ask are always the simplest, least sophisticated questions: Did the test prove or disprove the statement? Did it work? Once you have answered these questions in your assignment, you can elaborate on your conclusions by providing evidence from the experiments and, if required, from the scientific literature.

More advanced scientific thinking happens when you move your questioning up a level and query the nature and components of the hypothesis. By asking questions, you will find it easier not only to think through a problem but to get downright creative. You will also find it easier to write a longer analysis, as each question requires an answer. Bit by bit, you will create a substantial critical appraisal of the subject.

Consider this simple example. In biology, a typical laboratory assignment may involve students pollinating two types of flowers: some that have just opened and some that have been open for two days. The students are testing the hypothesis that newly opened flowers are more likely to be pollinated and to set seed than are older flowers. The students—behaving as bee surrogates—dust on the pollen. After a number of weeks, the students count the seeds per flower and analyze the data.

The first question to be answered is the easiest, as it comes directly out of the hypothesis: Did the newly opened flowers pollinate more readily and produce more seeds? Then the student can ask other questions by picking away at the hypothesis: Were the newly opened flowers already pollinated while in the bud? Were the older flowers diseased? Are bees more selective than their human surrogates, choosing to pollinate only newer flowers in nature? These questions cast doubt on some of the assumptions of the experiment. At this point a good student might suggest how to design a better experiment. Another way to identify similar shortcomings is to ask the question, "What is the matter with this experiment?" Following up on this line of thinking stirs up good questions. Oddly enough, negative thinking—that is, asking what's *wrong*—is much more effective than asking what's right. In science, this is almost always a winning strategy.

It is important to understand the difference between a *hypothesis* and a *thesis* (the difference between the two is described in greater detail in Chapter 2). A hypothesis is a statement related to an experiment—for example, "Are newly opened flowers more likely to be pollinated and set seed than older flowers?" On the other hand, a thesis is a statement related to an essay—for example, "Bee populations around the world are threatened by extensive use of agrochemicals." A thesis is also the name given to a dissertation, which is the scholarly work that summarizes research conducted, for example, by an undergraduate student upon completion of an honours program, or by a graduate student upon completion of her or his doctoral research.

Guidelines for Writing

Whenever you embark on a writing project, try to keep the following guidelines in mind:

- Think about the reader(s) of your writing.
- Be clear about your subject and your purpose, what it is you hope to achieve.

- Define your terms.
- Include only relevant material; don't pad your writing to achieve a certain number of words.
- Strive for consistency of expression throughout the work.
- Make sure you are accurate in all of your statements, in your analysis and presentation of data, and in your documentation of sources.
- Order your information logically.
- Be simple and clear in expressing your ideas.
- Make sure that your argument is coherent.
- Draw conclusions that are clearly based on your evidence.
- Allow yourself lots of time to work on drafts before completing the final copy.
- Make sure to edit and proofread your work carefully.

In the chapters that follow we will consider these guidelines in greater detail.

Using Bias-Free Language

Just as you give thought to your reader and to the kind of tone you wish to create, you will also want to take pains to use language that steers clear of any suggestion of bias. The potential for bias is far-reaching, involving factors that include gender, race, culture, age, disability, occupation, religion, and socio-economic status. An unfortunate consequence of separating students into different faculties is that young scientists, particularly in their first few years of university or college, may be unaware of how deeply life sciences need to be free of bias. Developing an awareness of sensitive issues will help you to avoid using biased language. To avoid bias in your writing, hew to the following guidelines.

Gender

At one time, it would have been acceptable to refer to a person of either sex as "he," a practice still preferred by some traditionalists:

> For an ecologist to study bears in the wild, he will need tracking devices.

But as sensitivity to sexist language has increased, we have become more careful about this use of a generic pronoun. Here are some options for avoiding the problem:

- Use the passive voice:

    Tracking devices will be needed by an ecologist studying bears in the wild.

- Restructure the sentence:

    An ecologist who studies bears in the wild will need tracking devices.

- Use "he or she," although this is cumbersome and should be used sparingly:

    For an ecologist to study bears in the wild, he or she will need tracking devices.

- Use the plural form:

    Ecologists studying wild bears will need tracking devices.

Another option is to alternate, using the masculine form in one instance and the feminine form in the next.

Recently, some writers have used the neutral "they" to refer to a singular antecedent:

> When a writer aims for simplicity, they will occasionally break with grammatical convention.

Be careful, however. This may still raise the hackles of a traditional reader because it is grammatically incorrect.

Race and culture

The names used to describe someone's racial or cultural identity often carry negative connotations for some readers although they are acceptable to others. For example, consider the term *Negro*. The search for neutral language has produced alternatives such as *black* and *African Canadian*, but no single term has gained universal approval. We have similar problems with the term *Indian*, with alternatives such as *Indigenous, Aboriginal, Native*, and *First Nations*. The best approach is to find out what term the racial or cultural group in question prefers. You should also remember that concepts of race differ according to context even within a university or a college, with social scientists, philosophers, and students themselves each using quite different points of reference.

The impact of genomics on life sciences has raised the issue of race, with the result that great sensitivity is required, especially when discussing current hot topics such as personalized medicine, race-targeted pharmaceuticals, and human evolution. Today, racial terms are often used in a slightly bewildering fashion, serving as substitutes for descriptions of undefined genetic characteristics. When discussing genetics, avoid generalizations about race by using concrete, scientifically valid descriptions.

There are many other areas where the effort to develop and use neutral language has made an impact. To be politically correct, writers usually refer not to *old people* but to *seniors* and to someone as *having special needs* rather than *being handicapped*. Whatever the situation, be sensitive to the power of the words you use and take the time to search for language that is bias-free but not cumbersome. Because words describing particular groups of people may change, the onus is on you to check. A simple test is to imagine what a person in a targeted group would think of the terminology you have used and the description you have written. If you are not sure, look into it and get the terminology straightened out satisfactorily.

Summary

Writing for the life sciences is generally formal in style. Outside of the usual requirements of restraining use of the first person and avoiding slang and contractions, science-related writing demands a slightly different tone—life sciences writers generally affect a neutral, impersonal tone. When proving or disproving hypotheses, you can enrich your analytical writing by developing a framework around questions that naturally arise from the work. This is where careful thought to the purpose of the writing pays dividends. The time you spend considering the assignment's value, length, and readership will also help you to achieve clarity in your writing.

2 Planning an Essay

Objectives

- Planning an essay
- Establishing the type of writer you are
- Structuring your essay
- Differentiating between descriptive essays and argumentative essays
- Working from the core idea—the thesis
- Knowing the difference between a thesis and a hypothesis
- Using the three-C approach
- Creating an outline

Planning and organizing an essay not only makes the task of writing easier but also improves the result. This is not a one-size-fits-all process, as the amount of time you spend on each stage will depend on the nature of the assignment. For a short, straightforward essay requiring little research, you will likely spend most of the time drafting and editing. For a more complex essay in which you must take into account what others before you have said, it's likely that over half of your time will be spent on research and planning. Understanding the total process of completing a written assignment will serve you well throughout your years of study.

The Planning Stage

Read the instructions

When instructors assign essays or lab reports, they generally provide written instructions. Read these carefully and understand what is being asked of you in a particular assignment. Once you start working on an assignment, it is easy to get distracted by the finer details of the writing process and to lose track of the many project components. Please refer to the checklist inside the front cover of this book; you will notice the last item is to check that you have included everything demanded of you by the instructor. It's good to periodically remind yourself of these tasks and goals.

Know the pros and cons of planning

Many writers believe that the planning stage is the most important part of the whole writing process. Certainly the evidence shows that poor planning usually leads to disorganized writing. In the majority of students' essays the single greatest improvement would be not better research or better grammar but better organization.

Some students claim they can write essays without any planning at all, driven entirely by enthusiasm. On the rare occasions when they succeed in creating a well-written essay, their process is usually not as spontaneous as they think. In most cases they have thought or talked a good deal about the subject in advance and have come to the task with fully formed ideas. More often, students who try to write a lengthy essay without planning just end up frustrated. They get stuck in the middle and don't know how to finish, or they suddenly realize that they're rambling.

Organizing your thoughts before you start to write has its immediate advantages. As with any complex task, the more you plan your strategy, the better your results. Once you have planned a few essays and developed a system for organizing your ideas before you write, you will find that you can write any essay—even one on a subject that you simply don't like—without difficulty. This is a powerful position to be in.

Yet planning can get in the way. Some students prefer to write essays that evolve slowly, maturing and changing along the way. They need to warm to the task with a bit of exploratory writing (see pages 50–51). This style can be summed up in one phrase—*getting something down on paper so that the rest will follow*. It is, nevertheless, absolutely critical to develop a plan as you progress with your writing.

Whether you organize before or after you begin to write, at some point you need to plan. It is worth remembering that well-planned essays get much better grades than unplanned ones.

Read primary material

Primary material is the direct evidence—usually journal articles, sometimes books—on which you will base your essay. Surprising as it may seem, the best way to begin working with this material is to give it a fast initial skim. If you are consulting a book, don't just start reading from cover to cover. First, look at the table of contents, scan the index, and read the preface or introduction to get a sense of the author's purpose and plan. In the case of a journal article, skim the abstract, the introduction, the results, and the opening paragraph of the discussion; skip the sections on materials and methods. Getting an overview will allow you to focus your questions for a more purposeful and analytic second reading. Make no mistake: a superficial reading is not all you need. You still have to work through the material carefully a second time. But an initial skim followed by a focused second reading will give you a much more thorough understanding than one slow plod ever will.

Secondary sources

In the life sciences, secondary sources are analyses of the primary material. While primary literature is the literature of invention and discovery, secondary literature is the literature of summation, discussion, and commentary. The first is peer-reviewed, meaning the articles have been vetted by experts, whereas secondary sources are not. Secondary sources include textbooks, reviews, encyclopedias, government reports, and other similar sources. Instructors and professors give *much* more weight to primary sources. In some disciplines, instructors discourage secondary reading in introductory courses. They know that students who turn to commentaries may be so overwhelmed by the weight of authority that they will rely too heavily on them.

In some instances—and especially in the sciences—your instructor will encourage you to review recent secondary literature on your chosen topic to see where your views stand in relation to those of the experts in the field. However, if you turn to commentaries as a way around the difficulty of understanding the primary source, or if you base your argument solely on the interpretations of others, you may end up producing a trite, second-hand essay. Your interpretation could even be downright wrong, because at this stage you might not know enough about a subject to be able to accurately evaluate the commentary.

Always be sure you have a firm grasp of the primary material before you turn to secondary sources. Secondary sources are an important part of learning and are essential to many research papers, but they can never substitute for your own active reading of the primary material. Once you get a feel for the subject, you should never be afraid of holding and developing your own views.

Analyze your subject

Whether the subject you start with is one that has been assigned or suggested by your instructor or is one that you have chosen yourself, it is bound to be too broad for an essay topic. You will have to analyze your subject in order to find a way of limiting it.

Ask questions

How do you form useful questions? Journalists approach their stories through a six-question formula: *who? what? where? when? why?* and *how?* For example, starting with the question *what?* and applying it to a biological problem, you might ask, "What kind of organism is it?"; "What makes study of this organism interesting?"; "What are the biological problems associated with this group of organisms?"

In the life sciences, which rely heavily on experimental and descriptive investigation, *how* questions form the most common line of enquiry. *How?* leads to interpretation and analysis, whereas *why?* can mislead students into ascribing purpose and design where there may be none. But you should also embrace *why* questions, as they often serve a broader purpose. *Why* questions range from simple questions such as "Why were these organisms chosen?" to deeper speculative questions such as "Why did this behaviour evolve among these organisms?" Questions such as "Who are the experts on this organism?" are very useful and can lead you quickly to discover the more important facts, theories, and controversies. The life sciences are built on expert evidence enshrined in journals, so asking *who?* is always very useful.

To take another subject, consider some of the same kinds of questions you could ask about the effect of global climate change on polar bears in the Canadian Arctic:

- *What* are the likely effects of ice loss due to global warming on the polar bear's ability to hunt and to reproduce? *What* are the mathematical models that predict polar bear populations in various climate change scenarios?
- *Where* are the polar bears found? *Why* is the distribution not random throughout the North?

- *Who* are the experts on polar bear population biology? *Why* are there so few?
- *Why* is there no intergovernmental organization on arctic mammals?
- *How* will polar bears be affected by a 10 per cent loss of spring sea ice? *How* will polar bears be affected by a 50 per cent loss of sea ice?

Most often, the questions you ask initially—and the answers to them—will be general, but they will stimulate more specific questions that will help you refine your topic and develop a thesis statement.

Try the three-C approach
A more systematic scheme for analyzing a subject is the three-C approach. It forces you to look at a subject from three different perspectives, asking basic questions about *components*, *change*, and *context*.

Components:
- What parts or categories can the subject be broken down into?
- Can the main divisions be subdivided?

Change:
- What features have changed?
- Is there a trend?
- What caused the change?
- What are the results of the change?

Context:
- What is the larger issue surrounding the subject?
- In what tradition or school of thought does the subject belong?
- How is the subject similar to, and different from, related subjects?

What are the *components* of the subject?
This question forces you to break down the subject into smaller elements. It helps you avoid oversimplification and easy generalization.

Suppose that your assignment is to write an essay on tree conservation and climate change. If you consider the components, you might decide that you can split the subject into (1) tree species that are most likely to be affected by climate change and (2) the challenges of tree conservation in a changing global climate. Alternatively, you might divide it into (1) the relationship between trees and climate, (2) the diversity of phenotypic and genotypic responses of trees to changing climate, and (3) various models of expected climate change in relation to the distribution of keystone tree species. Because these are still broad topics, you might break them down further, for example, by splitting the relationship between trees and climate into (1) common forest trees that are likely to be vulnerable to changes, (2) common trees unlikely to be affected, (3) rare trees (e.g., red-listed species) that are particularly vulnerable, (4) rare trees that are expected to rebound and spread under altered conditions, and so on.

Similarly, if you were analyzing the signal transduction pathway in a stress response in yeast, you could ask, "What are the extracellular components?" (e.g., composition of the extracellular matrix, ligands) and "What are the intracellular components?" (e.g., membrane-bound proteins, enzymes, signalling molecules, transcription factors, metabolic responses, and genes). Or you could ask, "What are the content groupings?" (e.g., signalling molecules, proteins, and genes). If you were writing an essay on predator–prey relations in the tundra, you could ask, "What are the major predators and their prey?" and "Is there a mathematical relationship between predator populations and prey populations?"

Approaching your subject this way will help you appreciate its complexity and avoid making wide-sweeping generalizations that don't apply to all areas of the subject. In addition, asking questions about the components of your subject may help you find one aspect of it that is not too large for you to explore in detail.

What features of the subject suggest *change*?
This question helps you to think about trends. It can also point to antecedents or causes of an occurrence as well as to the likely results or implications of a change.

For the essay on tree conservation and climate change, you could ask, "What kinds of adaptations do trees have to climate change?" You might also ask, "What species are most vulnerable?"; "Are particular ecosystems more susceptible than others?"; or "What is known about temperature tolerances of different species of trees?"

Suppose you have decided to focus on stress-induced signal transduction pathways in yeast. You might ask, "What are the effects of changes in gene expression on stress tolerance?" or "Do proteins in this pathway change their physical conformation, resulting in an altered function?" If the experiment does not reveal or explain the complete pathway, you might also question how future research might change the general understanding of how pathways are controlled.

On the subject of predator–prey relations in the tundra, you might ask, "Have there been changes or trends in the population growth of predators that are not influenced by prey availability, such as disease, habitat alteration, or some human influence?" or "Have there been changes in the type or abundance of prey available?" Then ask, "What are the causes and reasons for such changes?"

What is the *context* of this subject?
This question forces you to see the bigger picture. What are the similarities and differences between your chosen subject and related ones? To what particular school of thought or tradition does the subject belong? Questions about context are *rarely* posed by first- and second-year students, largely because the general emphasis on rote learning during the first two years of study tends to create an impression that knowledge is static. Enquiring about the context will quickly reveal to students that many fields in the life sciences presently are—or have always been—in intellectual ferment, with theories hotly debated. The following are typical context questions:

- What are the different climate scenarios for Canadian biomes in 50 years? In 100 years? Can assisted migration be implemented for vulnerable tree species? Are the limits on implementing assisted migration industrial or governmental?
- How does stress-induced signal transduction compare with other types of signal transduction? Does the pathway in yeast differ from pathways found in other organisms?
- What theories exist to explain predator–prey relations? How do these theories apply to the organisms in your paper?

General as most of these questions are, you will find that they stimulate more specific questions—and thoughts—about the material from which you can choose your topic and decide on your controlling idea. Remember that the ability to ask intelligent questions is one of the most important, though often underrated, skills that you can develop for any work, in school, and elsewhere.[1]

Analyze a prescribed topic

You need to analyze prescribed or assigned topics carefully. Try underlining key words to make sure that you don't neglect anything. Distinguish the main focus from subordinate concerns. A common error in dealing with prescribed topics is to emphasize one portion while giving short shrift to another. Give each part its proper due, and make sure that you actually do what the instructions tell you to do. For example, consider the instructions implicit in these verbs:

Outline   State simply, without much development of each point.

Trace   Review by looking back—on stages or steps in a process, or on causes of an occurrence.

Explain   Show how or why something happens.

Discuss   Examine or analyze in an orderly way. This instruction allows you considerable freedom, as long as you take into account contrary evidence or ideas.

Compare   Examine differences as well as similarities. (For a more detailed discussion, see pages 53–54).

Evaluate   Analyze strengths and weaknesses, providing an overall assessment of worth.

These and other verbs tell you how to approach the topic; don't confuse them.

Develop a thesis

Every essay needs a controlling idea around which all the material can be organized. This central idea is usually known as a thesis, though in the case of an expository essay you may prefer to think of it as a theme. Consider these statements:

THEME: Viruses play a role in plant extinction.

THESIS: Agricultural practices in North America are contributing to the spread of plant viruses and a loss of plant diversity

The first is a straightforward statement of fact; an essay centred on such a theme would probably focus on simply providing examples of viruses that have caused extinction or near-extinction of plant species. By contrast, the second statement is one with which other people might well disagree; an essay based on this thesis would have to present a convincing argument. The expository form can produce an informative and interesting essay, but many students prefer the argumentative approach because it's easier to organize and is more likely to produce strong writing.

If you have decided to present an argument, you will probably want to create a working thesis as the focal point around which you can start organizing your material. This working thesis doesn't have to be final: you are free to change it at any stage in your planning. It simply serves as a linchpin, holding together your information and ideas as you organize. It will help you define your intentions, make your research more selective, and focus your essay.

At some point in the writing process you will probably want to make your working thesis into an explicit thesis statement that can appear in your introduction. This is not always necessary: as you gain experience, you may choose to present your thesis later or to imply it rather than state it. Even if you don't intend to include it in your final draft, though, you need to know what your thesis statement is and to keep it in mind throughout the writing process. It's worth taking the time to work this statement out carefully. Use a complete sentence to express it, and above all make sure that it is restricted, unified, and precise.[2]

Thesis versus hypothesis—what's the difference?
Science students should not confuse a *thesis* with a *hypothesis*. A thesis is a statement that needs to be proved and generally serves as an organizing principle for an essay or book. It relates directly to the central theme of a writing assignment. A hypothesis is the core idea of an experiment. In short, proving a thesis depends on rallying written fact and opinion, whereas testing a hypothesis depends on repeatable acquisition of data. A thesis is central to an essay; a hypothesis is central to a lab report. For information on how to formulate a hypothesis, see Chapter 5.

A restricted thesis
A restricted thesis is one that is narrow enough for you to examine thoroughly in the space you have available. Suppose, for example, that your general subject is evolution. Such a subject is much too broad to be handled properly in an essay of 1000 or 2000 words; you must restrict it in some way and create a line of argument for which you can supply adequate supporting evidence. Following the analytic questioning process, you might find that you want to restrict it to major influences on evolutionary thought: "The most important advances in evolutionary theory have come from population genetics." Or you might prefer to limit it to a discussion of key experiments: "Evolution is experimentally verifiable, as a number of key studies have shown."

As another example, suppose that your general subject for a 2000-word essay is the work of the famous nineteenth-century explorer and biologist Alexander von Humboldt. You might want to limit your essay by discussing a prominent theme in his journals: "Humboldt's personal comments reveal that he yearned to create a unified system that could explain the interrelatedness of the thousands of organisms he had seen." Or you could focus on some aspect of Humboldt's thought: "During his trip to South America, Humboldt developed an idealized perception of nature because he ignored the influence of humans on the biogeographic regions that he visited." Whatever the discipline or subject, make sure that your thesis is restricted enough that you can explore it in depth.

A unified thesis
A unified thesis must have one controlling idea, for example, "The aquaculture industry in the coastal waters of British Columbia is a growing enterprise." Beware of the double-headed thesis: "The aquaculture industry in the coastal waters of British Columbia is a growing enterprise, but its failure to address sea lice is an increasing concern." What is the controlling idea here? Is it the history of aquaculture or the concern about fish pathology? It's possible to have two or more related ideas in a thesis, but only if one of them is clearly in control, with all the other ideas subordinated to it: "Recent trends in capital investment have led to the growth of aquaculture in the coastal waters of British Columbia, but too little is being invested in studying biological diseases." In this example, the controlling idea is investment of capital in aquaculture.

A precise thesis
A precise thesis should not contain vague terms such as *interesting* and *significant*, as in "The government of Ontario's development of policy concerning water quality is one of the most interesting cases of environmental regulation." Does *interesting* mean "effective," "daring," "controversial," or "intriguing"? Don't say simply, "The role of honeybees in pollination is important" when you can be more precise about the role of bees in plant reproduction and the advantages that plants get out of their relationship with bees: "The numerous types of pollination that bees are able to perform improve the fitness of plants."

Remember to be as specific as possible when creating a thesis in order to focus your essay. Don't just make an assertion—give the main reason for it. Instead of saying, "Canadians are more accepting than Americans of evolution" and leaving it at that, add an explanation: "Canadians are more accepting than Americans of evolution because Canadian public schools and government agencies are less influenced by religious principles than are their American counterparts." If these details make your thesis sound awkward, don't worry; a working thesis is only a planning device, something to guide the organization of your ideas. You can change the wording in your final essay.

Create an outline

Individual writers differ in their need for a formal plan. Because organization is such a common problem, though, it's a good idea to know how to draw up an effective plan. Of course, the exact form your plan takes will depend on the pattern you use to develop your ideas—whether you are defining, classifying, or comparing, for example (see pages 51–54).

For most students, a well-organized outline in point form is the most useful model. It is important that the system that you choose allows you to see, *at a glance*, the overall outline of your work (see *Code your categories*, on page 24).

The following is an example of an outline for an argumentative essay:

> THESIS: Trophy hunting in British Columbia is an unsustainable elitist sport whose origins and regulation are not in wildlife management, as claimed by the government of British Columbia, but in the long-discredited practice of placing a bounty on wildlife.

    I. Trophy hunting in British Columbia
        A. Trophy species
            1. Current species
            2. Former species
        B. Effect of hunting on population structure of trophy species
            1. Historical populations, quotas, and policies
            2. Current populations, quotas, and policies
        C. Types of trophy hunters
            1. Hunter's citizenship
            2. Hunter's income
        D. Types of trophy hunting
            1. In natural settings
            2. In restricted settings
        E. Trophy hunting policies
            1. BC versus the rest of Canada
                a) With similar species
                b) With different species

2. BC versus the rest of the world
   a) With similar species
   b) With different species

II. Bounty hunting
   A. Bounty species—wolves and bears
      1. History of wolf and bear bounties in Canada
      2. Comparison to history of wolf and bear bounties in the US
   B. Effect of bounty on species populations
      1. In Canada: ineffectual
      2. In the US: devastating
   C. Elimination of the bounty in Canada
      1. Opposition to bounty
      2. Legislative response for wildlife management
   D. Trophy hunting in BC
      1. Government justification
      2. Sporting association justification
      3. Opposition
   E. Today's trophy hunters are tomorrow's bounty hunters
      1. Interest in new forms of bounty hunting
      2. Goals of hunters' lobby groups

Conclusion

The following example shows an outline for an expository essay with a more descriptive theme:

THEME: Because of their importance for survival, refined navigational skills are found throughout the animal kingdom. These skills are based on a wide variety of different cues.

I. Orientation and navigation in different kinds of animals
   A. Birds
      1. Arctic tern
      2. Homing pigeon
   B. Insects
      1. Army ant
      2. Monarch butterfly

    C. Fish
        1. Salmon
        2. Eels
        3. Tuna

II. Navigating serves different purposes
    A. Migration
        1. Favourable climate
            a) Going south
            b) Going north
        2. Food availability
        3. Breeding grounds
    B. Locating local food sources
        1. Honeybee
        2. Ant

III. Navigating animals use different cues
    A. Celestial cues
        1. Sun compass
        2. Star navigation
    B. Terrestrial cues
        1. Geomagnetism
        2. Barometric pressure
        3. Odour trails
        4. Landmarks
    C. Neurobiology of cue interpretation
        1. Eye and brain
        2. Antennae

Conclusion

The guidelines for both types of outline are simple:

- **Code your categories**. Use different sets of markings to establish the relative importance of your entries. The examples here move from roman numerals to uppercase letters to Arabic numerals to lower-case letters, but you could use a different system. When typing your outline

in a word processor, you can select from various list styles to organize the different heading levels.
- **Categorize according to importance**. Make sure that only items of equal value are put in equivalent categories. Give major points more weight than minor ones.
- **Check lines of connection**. Make sure that each of the main categories is directly linked to the central thesis, and then see that each subcategory is directly linked to the larger category that contains it. Checking these lines of connection is the best way of preventing essay muddle.
- **Be consistent**. In arranging your points, use the same order every time. You may choose to move from the most important point to the least important, or vice versa, as long as you are consistent.
- **Be logical**. In addition to checking for lines of connection and organizational consistency, make sure that the overall development of your work is logical. Does each heading/idea/discussion flow into the next, leading your reader through the material in the most logical manner?
- **Use parallel wording**. Phrasing each entry in a similar way makes it easier for your reader to follow your line of thinking. For a discussion of parallel structure, see pages 137 and 138.

Summary

Writing essays in the life sciences can be a creative and imaginative experience. This process is easier if you have a well-organized set of hooks on to which you can hang information. As you develop your writing abilities and organizational style, particularly in the first years of university or college, take time to explore the different methods for structuring an essay provided in this chapter. Whatever style you choose, organizing your ideas and posing the questions that will drive your essay—whether it's descriptive or argumentative—are important first steps.

Notes

1. For a more detailed discussion of heuristic procedures, see Richard E. Young, Alton L. Becker, and Kenneth Pike, *Rhetoric, Discovery and Change* (New York, NY: Harcourt Brace Jovanovich, 1970), 119–36.
2. Joseph F. Trimmer, *Writing with a Purpose*, 12th ed. (Boston, MA: Houghton Mifflin, 1998), 62–3.

3 Researching an Essay

Objectives
- Using library resources
- Searching online databases
- Assessing electronic resources—the Web
- Critical thinking when incorporating a source into one's own writing

If your topic requires more facts or evidence than the primary material provides, or if you want to know other people's opinions on the subject, you will need to do some research. Some students like to read around in the subject area before they decide on an essay topic; for them, the thesis comes after the exploration. You may find this approach useful for some essays, but generally it's better to narrow your scope and plan a tentative thesis before you turn to secondary sources—you'll save time and produce a more original essay.

Explore Library Resources

Before you begin your first research assignment, get to know your way around your school's library. This step is essential. You don't want to be so overwhelmed by the library's size and complexity that you either scrimp on required research or waste time and energy trying to find information. Most academic libraries have orientation seminars specifically designed to show you where and how to find what you want. Take advantage of these services. Librarians will be glad to show you the bibliographies, indices, online databases, and other reference tools for your field of study. Once you are familiar with these basic resources, you will be able to check systematically for available material.

In addition to on-site resources, libraries offer a wealth of online services that you can access from remote locations. These services allow you to conduct your initial search for materials from your own home computer.

Online catalogues provide remote access to a list of all the holdings at your library, including books, videos, microforms, and print and online journals. A search by keyword will give you a list of relevant sources in your library and possibly in other libraries as well. Interlibrary loan services enable you to access these off-campus resources quickly at little or no cost. Give yourself lots of time to conduct your preliminary online search. Even if you have a good idea of what's available from the online catalogue, you will still need to go to the library to verify that the content of the shelf holdings matches the content described in the online catalogue. If you find that your library's resources are not extensive enough for what you need, you should consider interlibrary loans, which are fast, but only fast *enough* if you plan in advance.

Electronic databases simplify your search for information because they make millions of journal articles available from a single source. Libraries subscribe to online database services such as Web of Science, JSTOR, PubMed, and EBSCOhost that index a subset of smaller databases, thereby acting as gateways to a huge network of online journals. A single search gives you access to articles in thousands of different journals.

To conduct a database search from a remote computer, simply go to your library's website and follow the link to the database service of your choice. You can then search by subject, author, or title. In addition, there are usually options for narrowing the search, for example by restricting it to a specific journal, date of publication, or discipline.

Your search results will provide you with a list of articles on your subject, some of which are available in full text; this format allows you to read the material online as well as save, print, or email it.

Browse Online Databases

How to use Web of Science, PubMed, and other online database services

Most libraries subscribe to online database services that allow you to access articles published in international peer-reviewed journals. Web of Science, PubMed, and JSTOR are large database services that allow you to search multiple databases at once, but there are also a host of useful discipline-specific databases—for example, CAB Abstracts, which provides material related to applied life sciences.

- **Web of Science** is suitable for any life sciences or health sciences essay, and it is fairly easy to navigate.
- **PubMed** is narrower in focus, restricting itself mainly to medicine and peripherally related subjects.
- **JSTOR** is an interdisciplinary resource, focusing generally on the humanities, social sciences, and sciences; it doesn't always provide the most recently published articles, but its archive contains many complete collections, some dating back to the nineteenth century. These older articles can be very valuable if you are looking at the history of an idea or field.

Do *not* confuse database services with Internet search engines. Google, Yahoo, Bing, and similar search engines find links to content published on the internet; PubMed and Web of Science find links to primary literature published in peer-reviewed journals. You are after peer-reviewed primary literature, so you must master these database services.

Google searches can be much improved if you use their free web search engine called Google Scholar. You can sign in if you have a Gmail account. Google Scholar has many sophisticated functions that allow you to find papers. One aspect of Google Scholar that is both good and bad is its broad coverage, which extends well beyond peer-reviewed literature. This is good, because you can find things not found in peer-review-oriented databases, but it is bad because it requires that you already understand the differences in quality between the various types of reports and articles. At the beginning of a science career this is a tall order.

You can effectively navigate databases if you realize one thing: the value of database services resides not in the amount of information they can retrieve—which is truly overwhelming—but in the ease with which they allow you to link papers through time. You can search backwards or forwards; you can often find the first paper published in a field; and, in a click or two, you can find the most recent publication in that same field. Database services also allow you to link papers based on other factors, such as citation rate (e.g., to find the most important papers) or place of origin (e.g., to find papers written by scientists in a particular laboratory).

Let's look at three different case studies that illustrate how to search a database for useful material. Web of Science will be used for these examples.

Case 1: Researching a scientist's body of work
Imagine that you are writing an essay entitled "David Schindler's Contribution to Aquatic Ecology." Professor Schindler, of the University of Alberta, is one of the world's foremost experts on freshwater biology. How do you know he's a world expert? How would you write a retrospective on this Canadian's contributions to the field of aquatic biology? As you plan your approach, remember that scientists produce science, and you can use their international peer-reviewed papers as currency to assess the value of their work.

You can start by visiting Google and searching "David Schindler, University of Alberta." His academic website with the University of Alberta provides a limited list of his publications. The valuable piece of information is that his publication name includes his complete initials. He is D.W. Schindler. With this information you can begin your search.

Web of Science is the portal for a number of databases, including a number that are very relevant to the life sciences: Web of Science Core Collection, MEDLINE, and Zoological Record. Once you've signed in through your library, you will see that you are on a page tabbed "Web of Science."

To begin your search for papers written by David Schindler, follow the steps below:

- First, click on the tab to the right of the Search arrow. It is a drop down menu that provides you with a variety of data base choices. Choose "All Databases."
- Move your cursor to "Topic," click on the arrow to the right, and a drop-down menu will appear. Select "Author." Type "Schindler" into the search field, and press the blue "Search" button on the right. This search will find more than 16,000 papers.
- Next, start a new search by pressing on the orange "Search" button at the top left-hand side of the page. Note the arrow points backwards, indicating that if you press on it you will go back a screen. Type "Schindler, D" into the "Author" field, and then press "Search." Your results will drop to roughly 1100.

As you look down the list of results, you will see a number of papers that have nothing to do with aquatic ecology. These search results are not

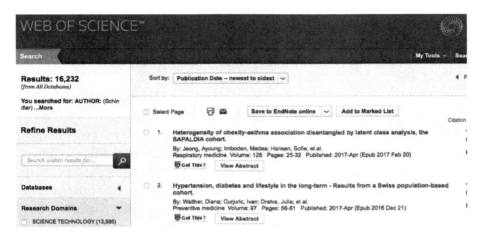

just one person's publication list, as there are many Schindlers with first names beginning with "D," including Damaris, Dirk, and Daniel. This is why we recommended that you first look up David Schindler's website with the University of Alberta, as you found out from that search that he publishes as D.W. Schindler.

- Now, press the "Search" button to return to the search page.
- Add the letter "W" so that it reads "Schindler, DW" in the "Author" search field. Click on the blue "Search" button.

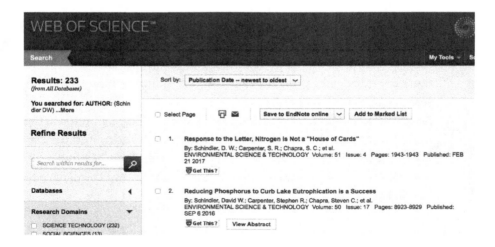

The result will be approximately 230 papers on aquatic ecology, all written by David W. Schindler. (This format, last name followed by initials, works for author searches in most databases.)

A few timely comparisons are in order at this point. The first comparison is between "All Databases" and "Web of Science Core Collection."

- Click on the "Search" tab at the top of the page; then click on the "All Databases" drop-down menu and select "Web of Science Core Collection." Then check that "Author" is still selected from the drop-down menu to the right of the search field, and type in "Schindler, DW" into the "Author" search field (it is not case sensitive). Click on the blue "Search" button.

The number of papers drops to around 210. Why the difference? "All Databases" includes more than "Web of Science Core Collection," but when you use "All Databases" you have less control of the output.

- Look on the left-hand side of the page at the light-grey box entitled "Refine Results." Note the categories included in "Web of Science Categories."
- Click on the blue triangle to open the category, if it isn't already open. If you repeat the search using "All Databases," the five research areas (e.g., "Fisheries," "Marine Freshwater Biology," etc.) would not display. In short, using different databases within Web of Science changes how you can search later on.
- Now look at the figure below this paragraph. Under the "Refine Results" box, the "Web of Science Categories" indicates the number of papers available in each category. This is a very useful feature as you will see later.

Let's return to David W. Schindler's publications.

Is it possible for a scientist to produce so many papers? Yes. Some scientists produce more than 300 during their careers. The value of the papers lies not in their number but in their quality. Good papers present useful information that other scientists will want to read and then cite in their own papers. The default in Web of Science is to present the papers in chronological order, with the most recent papers first.

- Now, look at the top of the "Results" page and you will find a "Sort by" drop-down menu, which is on the setting "Publication Date—newest to oldest." If you look down this first page of results, you will note that recently published papers have low citation numbers—refer to "Times Cited," located on the right side of the page.
- Next, let's take a look at Schindler's early career. If you look to the top right you will see that you are viewing page one of more than 20 pages of citations.
- Click on the "Sort by" drop-down menu and choose "Publication Date—oldest to newest." This arranges the list in the reverse order.

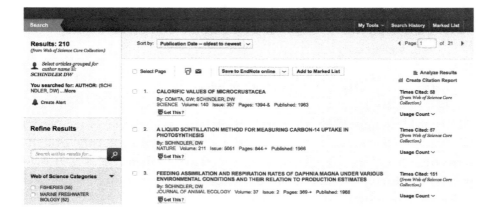

As you would expect from a world-leading scientist, Schindler's papers have extraordinarily high citation rates.

- Click on the "Sort by" drop-down menu again, but this time choose "Times Cited—highest to lowest." One paper by Schindler was cited more than 2600 times by other scientists! The average citation rate for the typical peer-reviewed paper is about one time, with many papers never cited at all. David Schindler's output is extraordinary by two measures that you have used: the number of publications and the importance (citation rate). His papers are among the most cited in Canadian scientific publishing history.

Web of Science gives you choices on how to search. There is a different way to carry out this author search, which you may prefer:

- Make sure that you are still in the "Web of Science" tab, and then click on the white arrow in the blue square to the right of the "Basic Search" heading. Choose "Author Search." The screen will refresh.
- Follow the prompts: type "Schindler" in the "Family Name" field, "DW" in the "Initials" field, and click on "Exact Matches Only."
- Then click on "Select Research Domain." You will be given four choices; click in the box to the left of "Life Sciences Biomedicine." Avoid the "Select Organization" button, as it does not provide information that will be useful to you. Press the blue button on the bottom-right side of the page entitled "Finish Search." Again, you get just over 200 papers.

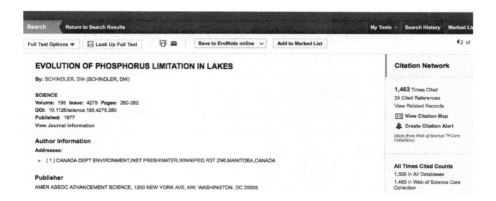

- Next, reorder the publication list by clicking on the "Sort by" menu and selecting "Times Cited—highest to lowest." Find the second paper, which is from 1977 and is entitled "Evolution of Phosphorus Limitation in Lakes." Click on the title. A new window that is dedicated to this article will appear.

The record entry for the article can be parsed by field.

- The title is followed by the author, D.W. Schindler. The paper was published in 1977 in *Science*, volume 195, issue 4275, on pages 260–262. The DOI number—10.1126/science.195.4275.260—is the digital object identifier that is used to link to this article.
- On the right side of the page it states that this paper has been cited over 1,400 times. The text is blue, which means it is linked and can be clicked. If you click on "Times Cited" you can view over 1400 references. The paper itself cited only 24 other papers. If you want to look at that list of papers, you can click on the blue 24.
- If you look on the right side of the record entry page, below "Times Cited" you will find another blue link that directs you to "View Citation Map"—this is a quite sophisticated way to look for other information, so you should avoid this feature until your upper years of study. Briefly, citation maps allow you to search forwards and backwards from any given reference. Forward searches identify research articles that have cited the article that you have in front of you, whereas backward searches identify the sources of articles. You need to have Java enabled for this to work. We recommend that you first try this with an article that has just a handful of citations.
- Of the remaining fields, the only useful one is "Addresses," which tells you where the authors work. You now know from this information that David Schindler worked in Winnipeg before moving to Edmonton.
- You can, at any point, send this information to your personal email account by clicking on the email icon, which is a blue envelope located above the article's title. Sending the article details to your email address is one way to keep track of papers that you might want to revisit or look up and read.
- Another way to keep track of papers is to create a marked list by clicking on the "Add to Marked List" button, which is located to the right

of the email icon. Marked articles can be exported into referencing software, such as EndNote, RefWorks, and others that are found on the drop-down menu.

So how do you access a paper? There are three ways:

- The first is to click on the "Full Text Options" menu, located on the top left-hand side of the page. Click on "Get This?"

   From here, you can download the entire article by clicking the "Download PDF" button, located on the right. In this example, JSTOR was probably not your only choice. Often, you can access journal archives with a single click. If not, it is relatively easy to go to your library's online catalogue and find a link to the *Science* online archives, from which you can directly access the article you want.

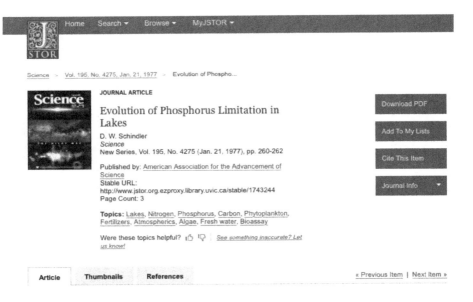

- The second way to access a paper is to click on the "Look Up Full Text" button, which directs you to Google Scholar. Once there, you will have links to the article. In this case it will be via JSTOR.
- The third way is old school: you may be forced to retrieve a printed copy from your library. You will need to use this method if your library does not have access to electronic back issues of *Science*. To begin, find

the journal in the online library catalogue and check that your library has a bound copy of the volume that you want. Write down the call number, volume, issue, and page information, then go to the stacks, get the volume, and photocopy the pages.

Once you've found your primary literature, how do you begin to evaluate the impact of David Schindler's work on the field of aquatic ecology? One way is to look at Schindler's articles that have had the greatest impact on other researchers.

- Return to Web of Science and your list of search results ordered by citation ("Times Cited—highest to lowest") by clicking on the field beside the orange "Search" button titled "Return to Search Results."

It turns out that there is another very fast way to find Schindler's most cited articles.

- Open a new browser window.
- Find Google Scholar and type in "Schindler DW"—you will get over 120,000 hits.
- Before you cry out "argh!" in frustration, look at the first entry—"Human Alteration of the Global Nitrogen Cycle: Sources and Consequences"—and look to the bottom left-hand corner of this entry, where you will see that it has been cited well over 5000 times. The next entry, "Evolution of Phosphorus Limitation in Lakes," is cited in more than 2300 other papers. Surely this is too easy! It's a list of Schindler's papers beginning with the most cited!

> **Google** schindler dw
>
> Scholar   About 120,000 results (0.08 sec)
>
> Articles   **Human alteration of the global nitrogen cycle: sources and consequences**
> Case law   ..., GE Likens, PA Matson, **DW Schindler**... - Ecological ..., 1997 - Wiley Online Library
> My library Abstract Nitrogen is a key element controlling the species composition, diversity, dynamics, and functioning of many terrestrial, freshwater, and marine ecosystems. Many of the original plant species living in these ecosystems are adapted to, and function optimally in, soils and
>   Cited by 5144   Related articles   All 24 versions   Cite   Save
>
> Any time   **Evolution of phosphorus limitation in lakes**
> Since 2017 **DW Schindler** - Science, 1977 - nature.berkeley.edu
> Since 2016 Your use of the JSTOR archive indicates your acceptance of JSTOR's Terms and Conditions
> Since 2013 of Use, available at http://www. jstor. org/about/terms. html. JSTOR's Terms and Conditions
> Custom range... of Use provides, in part, that unless you have obtained prior permission, you may not
>   Cited by 2337   Related articles   All 7 versions   Cite   Save   More
>
> Sort by relevance **Eutrophication of lakes cannot be controlled by reducing nitrogen input: results of**
> Sort by date   **a 37-year whole-ecosystem experiment**
>   **DW Schindler**, RE Hecky, DL Findlay... - Proceedings of the ..., 2008 - National Acad Sciences
> ✓ include patents Abstract Lake 227, a small lake in the Precambrian Shield at the Experimental Lakes Area
> ✓ include citations (ELA), has been fertilized for 37 years with constant annual inputs of phosphorus and
>   decreasing inputs of nitrogen to test the theory that controlling nitrogen inputs can control
>   Cited by 794   Related articles   All 20 versions   Cite   Save

On that last search, both Web of Science and Google Scholar were useful. This comparison illustrates a far deeper truth: we are all flicking and bouncing through information in ways that are not actually systematic but nevertheless work. Instead of digging deeper and deeper into a single source of information, we often skip horizontally through databases. How many databases will you have open on your computer when you're researching? Probably more than one.

How we research topics today is not how we conducted our searches 30 years ago. Indeed, your professors are doing just what you are doing—skimming more and reading less. Science articles are designed to be read quickly. Is this because we don't like reading as much as we did in the past? We don't think so. Skimming is a superficial but very efficient way of filtering information. We're busy and we want answers quickly, and Internet-based searches spare us a great deal of laborious hunting. Online searches can also surprise us with unexpected information, forcing us to quickly revise our search strategies. The modern method is actually more fun, turning research into an exciting, fast-paced hunt. When we compare electronic research methods with traditional scholarship over the centuries, we find that we're in the midst of a revolution in research behaviour.

Case 2: Researching intersecting subjects
Now imagine that you are writing an essay on proteomics and its contribution to our understanding of sleeping sickness. How do you find material on a

recently emerged, rapidly expanding field such as proteomics? How do you find relevant information on a topic such as sleeping sickness that researchers have been studying for over a hundred years?

It is best to remember that science advances along fronts, sometimes very rapidly. Usually, instructors and TAs are only interested in what's modern and topical. This means that you must restrict your search to very recent articles (e.g., the last five years).

Before starting this exercise, close and reopen Web of Science to restore it to its default settings.

- Begin a new search by first making sure that the heading beside the orange "Search" tab is set to the "Web of Science Core Collection."
- In the "Timespan" section, click on the button beside "From," and then select 2012 from the menu. The range should now read "From 2012 to 2017."

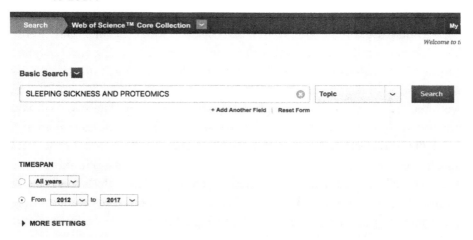

- To begin your search, type "sleeping sickness and proteomics" into the "Topic" field, and click on the blue "Search" button. Web of Science is not case-sensitive; you can use upper-case or lower-case letters.
- The screen that pops up indicates that there are about a dozen papers that meet your requirements.

But now you have a problem—you have so little choice and such a big essay to write. Surely there are more papers? One good way to find papers related to your topic is to look at the citations within one of these papers.

3 Researching an Essay | 39

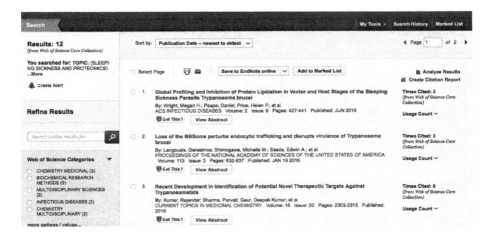

- Scroll down the list and select the paper by Holzmuller et al. (2013) by clicking on the title. Information including the title, authors, journal, and abstract will appear. Note that all matches to the search terms "sleeping sickness" and "proteomics" are highlighted.

- To reveal the list of articles referred to by the authors of this paper, click on "Cited References" (located below the "Citation Network" heading). The blue number "103" to the right of the "Cited References" heading indicates that the authors have cited 103 other papers. A list of those papers, arranged in alphabetical order, will pop up.

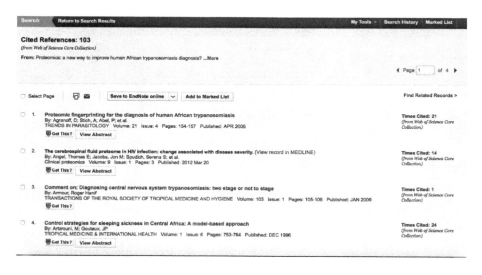

- This list reveals that similar studies, with slightly different titles, are available.
- As you scroll through the list of cited references, you will notice that many are not relevant, but a few look promising.
- A quick glance down the first of the five pages provides a number of on-topic papers. For example, there is a paper by Agranoff et al. (2005) entitled "Proteomic Fingerprinting for the Diagnosis of Human African Trypanosomiasis." The paper is older, but it is very well-cited (21 times), which means that it is highly thought of and that you ought to include it in your essay. These 21 citations will be, in turn, useful for finding other more modern papers.

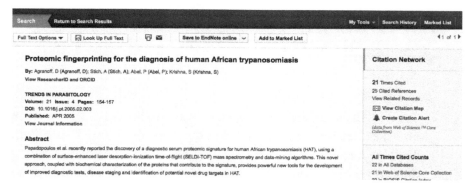

These research techniques can be very useful for finding relevant articles, and gaining familiarity with these tools now will substantially shorten the time you need to spend preparing for your next paper.

How to use search operators

It is essential that you understand the role of search operators in order to effectively broaden or narrow your search results. At the beginning of the second case study, you typed "sleeping sickness and proteomics" into the search field. "AND" is a type of Boolean search operator that controls the database search, limiting it to papers that are on both sleeping sickness *and* proteomics. There is an implicit "and" between the words *sleeping sickness*. Whether you type "sleeping and sickness and proteomics" or "sleeping sickness proteomics" into the "Topic" field, you get the same number of records in the "Results." However, if you type quotation marks around "proteomics sleeping sickness" you will get nothing as quotation marks in any field are used to search for the exact phrase. Because "proteomics sleeping sickness" as an exact phrase is nonsense, the result is zero matches.

If you search "proteomics OR sleeping sickness," a list of papers will be produced that contain any of the terms, including all papers on proteomics and all papers on sleeping sickness, which totals many tens of thousands of papers. The result is a deluge of information.

There are other search operators that you can use in the "Topic" field, including "NOT" and "NEAR/x." When searching in the "Address" field, you can also use the search operator "SAME."

- If you click on "Help" in the upper right-hand corner of the Web of Science page, the "Help" page pops up.
- In the upper right-hand corner are two links; choose "Contents" and then scroll down the page until you get to "Search Rules."
- Click on "Search Operators," and a much more detailed page pops up that has many examples set in a grey box along the right-hand side of the page.

Here is a quick overview on how to use search operators:

- Use "AND" to find records with all terms.
- Use "OR" to find records with any of the terms.

- Use "NOT" to exclude records with certain terms.
- Use "NEAR/x" to find records in which the terms appear within a certain number of words of each other—replace the *x* with a number to set the maximum number of words separating the terms (e.g., buffalo NEAR/10 migration).
- Use "SAME" when searching using the "Address" field to find records in which the terms appear in the same address (e.g., Laval SAME forestry).

There are also symbols that act as wildcards:

- Use the asterisk (*) for any group of characters, including no character (e.g., Schindler D* finds Schindler D, Schindler DW, Schindler David, and so on).
- Use the question mark (?) for one character (e.g., wom?n finds woman, women).
- Use the dollar sign ($) to represent zero or one character (e.g., behavio$r finds behaviour, behavior).

Parentheses allow more complex statements to be made:

- Use parentheses to group search terms—e.g., (conifers AND mature trees) and (micropropagation OR somatic embryogenesis).

Case 3: Researching a broad topic

Imagine that you're researching a very broad essay topic—what's new in CRISPR and human embryo research? This subject is tricky as it not only involves biology and medicine but it also has created significant political, philosophical, and ethical issues. How can you tackle such a vast subject? How do you, as an undergraduate student, find your way into this complex field and come out alive?

You may want to consult a few different sources before you turn to the primary literature. Reference librarians are very helpful. They can give you a basic overview and provide you with a quick insight into the various resources available at your institution. They will know if there are, for example, specialist reference collections, which may include online archives. Another source that may be useful is Wikipedia, which often contains useful articles on big subjects. Wikipedia relies on a self-editing community. The larger the community, the more vigilant it tends to be, especially on controversial topics. In the case of such a publicly discussed topic as CRISPR research, it can be an excellent entry point into the subject. However, a note of caution is required at this point. Wikipedia is made up of collaborative websites that allow *anyone* to contribute or modify content. A free encyclopedia that anyone can edit may allow you to find information quickly and easily, but it doesn't contain the kind of sound research you want to base your essay on. Although editors may continually check the veracity of contributions to the site, there is no guarantee that what you are reading is accurate. As a result, you should always check with your instructors whether they will accept Wikipedia as a reference source in your assignment. In formal science papers, it is generally unacceptable to cite Wikipedia because science is based on peer-reviewed literature, and Wikipedia is not peer-reviewed.

In addition, socially important scientific subjects not only create a wave of books and magazine articles, but there are also newspaper editorials that you can consult. News publications such as *The Globe and Mail*, *The New York Times*, *The New Yorker*, *The Guardian*, the *Independent*, and *The Times* often print editorials and articles that present pithy appraisals of controversial topics. You can also use Web of Science to find relevant articles published in generalist journals such as *Scientific American* and *New Scientist*, both of which excel in scientific journalism. Their articles tend to be very undergraduate-friendly, as they're written for non-specialists.

Once you have a general understanding of your topic, you will be ready to consult the primary literature.

- Begin a new search in Web of Science.
- In the drop-down menu beside the orange "Search" tab, select "All Databases."
- Searching "CRISPR and human embryo" among papers published in the last 10 years returns over 325 records.

This mountain of papers is still unmanageable, so you need to refine the search. Remember: you don't need to find all of the most technically advanced articles, just enough to form a reasonable review.

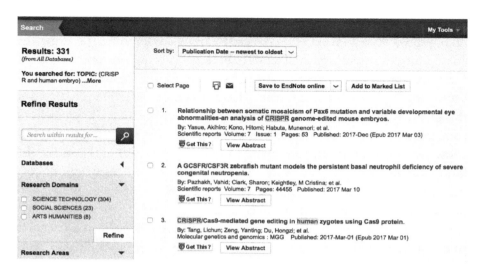

- Without leaving the page, look on the left-hand side where the "Refine Results" menu is found. Each category (e.g., "Databases") has a triangle located to the right of the heading.
- Scroll down to "Publication Years" and click on the triangle that drops down the menu. Select last year and the current year.

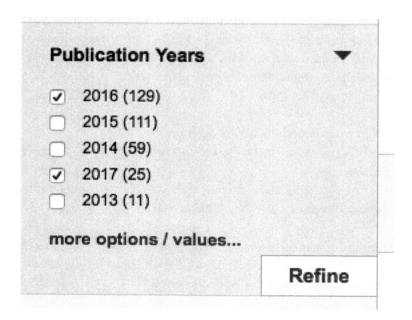

- Click on the "Refine" tab. The number of papers is now reduced to half.
- To find reviews, you will need to scroll up to "Document Types"; click on the triangle, which reveals the categories (e.g., "Article," "Review," "Editorial"), followed by "more options/values." Click on the latter. When the Web of Science page refreshes, you will see the different document types listed in order of abundance in the main viewing panel (i.e., documents classified as "Article" are more abundant than those classified as "Review," and so on).

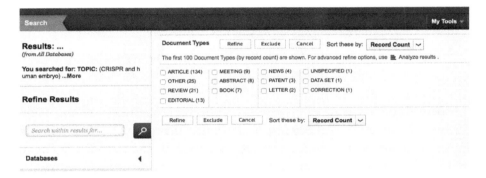

- To see a list of these reviews, check off the box beside "Review," and click the "Refine" button. A new list of a few dozen reviews will tumble out.
- You can further refine this list by looking in the grey side menu and clicking first on the triangle beside "Source Titles" and then on the new "more options/values" button that appears.

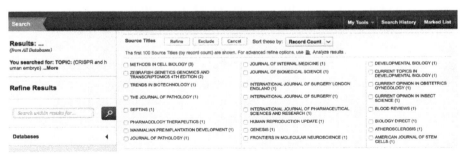

- Now you can choose which journals you'd like to consult by checking the boxes beside your selections. You could focus on journals that specialize in human embryos or medicine or you could restrict your search to journals that you know your library carries. Click the "Refine" button. You will now have a manageable list of articles.

Before you begin your first research project, play around with the many features offered by Web of Science. You will find that you can refine searches in very complex ways. You are not restricted to searching by topics, authors, and publication names, as the category buttons provide drop-down menus that allow you to select half a dozen other categories. For example, you could combine two topics—say, CRISPR and ethics—and find about 100 articles published over the last decade. In comparison, you could enter the same words, CRISPR and ethics , but instead of "Topic," select "Title" as the search category from the same drop-down menu, and you will find just two articles published in the same period. You can also find useful articles by searching for unusual terms within the text of the articles. Alternatively, you can use a cited reference as a starting point for a search, create citation maps that go forwards and backwards, and search through these maps—this method is very effective when you want to conduct a narrow search. With a bit of imagination coupled with experience, you can learn to find what you need quickly. Once you know how to use Web of Science, branch out and try other database services, such as PubMed. All are relatively easy to learn if you start with the help page.

Search the Web

In addition to the online services provided by your library, you may wish to access the huge volume of information available on the Internet. No doubt you already have favourite search engines and are accustomed to finding information on just about anything. Your search engine can access billions of sites—a wealth of information—almost instantly. Be careful, though. You will want to invest a little time and energy in making sure that the information you get is reliable and accurate. When writing essays for first- and second-year courses, you can use the Internet to help orient some of your questions, but you should rely only on peer-reviewed journal articles to support your points in an essay.

Evaluating online sources

Unlike academic journals, which are peer-reviewed and tend to be reliable sources of information, many websites do not have editorial boards and publish material that has not undergone any review process. Remember that anyone can publish online, as the current proliferation of blogs will attest. You want to be sure that the author or publisher of any material you use has the necessary authority to lend credibility to the site.

Consider the following tips when evaluating websites:

- **Determine the currency of the site.** Individual sites (and pages) should include a clear indication of when the material was written, published, and last revised.
- **Evaluate the accuracy of the information.** Be sure to cross-reference facts and figures against other sources. Data published on the site should be documented in citations or a bibliography, and research methods should be explained.
- **Be wary of blogs.** Although some companies have official blogs that can offer good advice about subjects such as cell biology, many are simply online diaries published by a rapidly increasing number of people who are expressing personal opinion and nothing more. Using such unverified material can seriously undermine your essay, as many a student has discovered.
- **Assess the overall quality of the site.** A major clue to the reliability of the site is its level of correctness and writing standard. Typos and grammatical errors are clear indications of unprofessional work that has not been monitored for correctness and accuracy. An online author who doesn't pay attention to these details probably doesn't have fastidious research methods either.

Use Critical Thinking

The answer to the question of which articles to include in an essay depends on the depth of your scholarship. Scholarship is learning of a high level. Students in their early years are at a disadvantage, because they are at a stage of building up their critical thinking. As you acquire and master the skills to judge scientific articles, your confidence increases. The more you write and the more you research subjects, the better your skills will become. It becomes easier to evaluate research, sources and even researchers. The problem for young scientists is that the tools available seem to be overwhelming. Take Web of Science. It is certainly a professional tool. There is nothing introductory or undergraduate-level about it.

How are you to find your way into this world of higher critical evaluation? How are you to make the correct choices? Two answers present themselves. One is to read until a subject becomes familiar, until themes start to repeat themselves, and patterns of thought become evident. If you read enough, thinking about a particular subject eventually becomes second nature. Another way to get to the same level is to work in labs and acquire research skills. Learning to conduct experiments and then to compare your results to experiments published in scientific journals provides you with the most important scientific experience you can have, which is ground-truthing of ideas with appropriate methodologies. Once you have lab experience, your relationship to scientific papers changes: these become more immediate documents. The experiences of other scientists become shared experiences, because working with similar organisms using similar experimental designs creates a sense of community. You realize that you are slowly but surely joining the working scientific community. You soon appreciate that peer-reviewed publications are the world of scientific enquiry based on repeatable experimentation and that some papers are far more informative than others. This directly improves your critical interpretation of primary literature. A balanced approach is one in which a young scientist reads passionately while conducting experiments in an established researcher's laboratory. This is the perfect way to develop critical faculties needed to become a successful scientist. If you get the opportunity, work in someone's lab. Faculty members and their graduate students will teach you how to evaluate literature.

To get a critical eye requires reading not only primary literature but reading reviews that discuss the directions and ideas that provide the momentum that propels a particular field of study. Every field has a history. Understanding that history turns a callow young scientist—a consumer of primary literature—into a discerning ambitious scientist.

Summary

Research in the life sciences involves conducting new experiments in the context of previous research. Consequently, learning to find peer-reviewed papers that are relevant to your essay topic is probably the most important step you will take in your development as a scholar of science. You must learn how to use library resources and retrieve relevant information from electronic databases. Practise searching for articles in Web of Science—if you can master such a powerful database, you will have the confidence to learn others. Once you develop these literature-searching techniques, you will be better prepared to enter the wider world of experimentation.

4 Writing an Essay

Objectives

- Developing your ideas
- Writing an effective introduction and conclusion
- Knowing the difference between a summary and a conclusion
- Using quotations
- Avoiding plagiarism
- Editing what you've created
- Learning by example—an annotated essay
- Backing up your work

Writing the First Draft

A first draft is not hard to assemble if you follow the suggestions outlined in this chapter. Normally, with a well-planned outline on hand, writing involves assembling the facts into a coherent whole. If you are good at writing, this is not a problem. However, students occasionally abandon their plans for various reasons. Here are some tips for students who struggle with writing the first draft.

Rather than striving for perfection from the moment they begin to write, most writers find it easier to compose the first draft as quickly as possible and do extensive revisions later. However you begin, you can't expect the first draft to be the final copy. Skilled writers know that revising is a necessary part of the writing process and that care taken with revisions makes the difference between a mediocre essay and a good one.

Many writers find it difficult to start writing. Facing down writer's block is not easy. The best thing to do is to look at the outline you've created and continue to work on the ideas, as developing them involves first writing down short phrases. This method has the virtue of creating the infrastructure of the essay, which will make it easier to expand your ideas into more detailed writing. If you are a student who suffers from writer's block, remember that you don't need to write all parts of the essay in the same order as they will appear in the final copy. For some, the introduction is the hardest part to write. If you face the first blank page with a growing sense of paralysis, try leaving the introduction until later and start with the first idea in your outline. If you feel so intimidated that you haven't even been able to draw up an outline, you might try the approach suggested by John Trimble[1] and begin anywhere: just write, "Well, it seems to me that . . ." and begin talking on paper. Instead of sharpening pencils or running out for a snack, just try to get going. Don't worry about grammar or wording; at this stage, the object is to get your writing juices flowing. Begin small: start linking your bits of evidence together to support your ideas, and then assemble these ideas into an argument.

Of course, you can't expect this kind of exploratory writing to resemble the first draft that follows an outline. You will probably need to do a great deal more changing and reorganizing, but at least you will have the relief of seeing words on a page. Experienced writers—and not only those with writer's block—consider this the most productive way to proceed when it's hard to get your ideas flowing. Once you get some words on the page, it is easier to build on this initial draft to develop your ideas.

Developing Your Ideas: Some Common Patterns

The way you develop your ideas will depend on your essay topic, and topics can vary enormously. Even so, most essays follow one or another of a handful of basic organizational patterns. Here are some of the patterns, along with suggestions for using them effectively.

Defining

Sometimes a whole essay is an extended definition, explaining the meaning of a term that is complicated, controversial, or simply important to your field of study—for example, *evolutionary biology* in medicine, *ethics* in animal biology, or *biotechnology* in agriculture. Rather than making your whole paper an extended definition, you may decide to begin your paper by defining a key term before shifting to a different organizational pattern. In either case, make your definition exact; it should be broad enough to include all the things that belong in the category but narrow enough to exclude things that don't belong. A good definition builds a kind of verbal fence around a term, herding together all the members of the class and cutting off all outsiders.

For any discussion of a term that goes beyond a bare definition, you should give concrete illustrations or examples. Depending on the nature of your essay, these could vary in length from one or two sentences to several paragraphs or even pages. If you are defining *biotechnology*, for instance, you will probably want to discuss at some length the theories of leading biotechnologists.

In an extended definition, it's also useful to point out the differences between the term you're defining and any others that may be related to or confused with it. For instance, if you're defining *parthenogenesis*, you might want to distinguish it from *cloning*; if you're defining *restoration ecology*, you might want to distinguish it from *conservation*; if you're defining *backcrossing*, you might want to distinguish it from *hybridization*.

Classifying

Classifying means dividing something into its separate parts according to a given principle of selection. The principle or criterion may vary. You could classify crops, for example, according to how they grow (above the ground or below the ground), how long they take to mature, or what climatic conditions they require. As another example, you could classify fauna found at a site according to species, population sizes of species, the niche or location that each species occupies, or taxonomic order. If you are organizing your essay by a system of classification, remember the following guidelines:

- You must account for all members of a class. If any are left over, you need to alter some categories or add more.
- You can divide categories into subcategories. You should consider using subcategories if there are significant differences within a category.
- You should include at least two items in each subcategory.

Explaining a process

Explanations of processes show how something works or has worked—whether it is the weather cycle, the Linnaean system of classification, or the stages in an insect's development. You need to be systematic, to break down the process into a series of steps or stages. Although the order of the steps will vary, most often it will be chronological, in which case you should verify that the sequence is both accurate and logical. Whatever the arrangement, you can generally make the process easier to follow if you start a new paragraph for each new stage.

Tracing causes or effects

A cause-or-effect analysis is really a particular kind of process discussion in which you explain how certain events have led to or resulted from other events. Usually you are explaining why something happened. When exploring causes and effects, always avoid oversimplifying relationships. If you are tracing causes, distinguish between a direct cause and a contributing cause, between what is a condition of something happening and what is merely a correlation or coincidence. There are many examples of spurious relationships: for instance, if you found that since 1980, carbon dioxide levels have increased and obesity levels have also increased, you cannot jump to the conclusion that increases in obesity cause increases in CO_2. Similarly, you must be sure that the result you identify is a genuine product of the event or action.

Comparing

Students sometimes forget that comparing things means showing differences as well as similarities—even if the instructions do not say "compare and contrast." Suppose, for instance, that you were comparing negative and positive opinions on the effects of conservation on the population biology of some rare animals. The easiest method for comparison—though not always the best—is to discuss the first subject in the comparison thoroughly and then move on to the second:

Positive views: population figures
conservation efforts
public opinion

Negative views: population figures
conservation efforts
public opinion

The problem with this kind of comparison is that it often sounds like two separate essays slapped together.

To make a successful comparison, you must integrate the two subjects, first in your introduction (by putting them both in a single context) and again in your conclusion (by bringing together the important points you have made about each subject). When discussing the second subject, try to refer repeatedly to your findings about the first. (For example, "In contrast to the rosy view that the government entertains of its conservation efforts, public opinion has recently swung behind conservationists who maintain that declining fish stocks point to poor management.") This method may be the wisest choice if the subjects you are comparing seem so different that it is hard to create similar categories by which to discuss them.

If you can find similar criteria or categories for discussing both subjects, however, the comparison will be more effective if you organize it by category:

    Population figures: negative views
                                    positive views

    Conservation efforts: negative views
                                    positive views

    Public opinion: negative views
                                    positive views

Because this kind of comparison is more tightly integrated, the reader will find it easier to see the similarities and differences between the subjects. As a result, the essay is likely to be more forceful.

Writing Introductions

The beginning of an essay has a dual purpose: to indicate your approach to the topic and to whet your reader's interest in what you have to say. There are many ways to catch your reader's attention; the following sections discuss four of the most common methods. However you choose to begin your essay, your lead must relate to your topic: never sacrifice relevance for originality. Also, whether your introduction is one paragraph or several, make sure that by the end of it your reader clearly knows the purpose of your essay and how you intend to accomplish it.

The funnel approach

One effective way of introducing a topic is to place it in a context—to supply a kind of backdrop that will put it in perspective. The idea is to step back and discuss the area into which your topic fits and then gradually narrow in on your specific topic. Sheridan Baker and Lawrence B. Gamache call this the *funnel approach*, where a broad statement at the beginning of your essay narrows to the argument that you explain and develop in the body of your essay.[2] For example, suppose that your topic is the growing demand for personalized medicine. You might begin with a more general discussion of the growing influence of DNA in crime-analysis procedures or in disease-screening techniques.

You can use a funnel opening in almost any kind of essay. The following example comes from an essay on phenotypic plasticity:

> Scientists once believed that one gene was responsible for one trait. Even today, popular language suggests that a single gene—for example, the "intelligence" gene or the "fat" gene—can control complex behaviour or morphology. However, more recently discovered complex interactions such as pleiotropy and epistasis prove that phenotypic traits are not a simple matter of genetics. Over the past decades, genetic researchers have increasingly focused on the occurrence and causes of phenotypic plasticity, an aspect of organismal life that merges genetics, anatomy, and physiology. Plasticity, in the developmental sense, is a property of the reaction norm for a specific expressed trait. Norms of reaction model the variability of phenotypes across different environments. Genes must interact with the environment, and that environment has a number of variable elements—temperature, climate, water availability, day length, soil nutrition, insects, and disease. Genes also interact with one another. When it comes to the biology of organisms, gene–environment interactions have far-reaching and often elusive consequences.

You should try to catch your reader's interest right from the start. You know from your own reading how a dull beginning can put you off a book or an article. The fact that your instructor must read on anyway makes no difference. If a reader has to get through thirty or forty similar essays, it's all the more important for yours to stand out.

The quotation

This approach works especially well when you use a quotation from the person or work that you will discuss in your essay. Here is an example from an essay on the effects of overfertilization on Canadian lakes:

> Lakes around dense human habitation are in danger of dying. This is an international problem, according to David Schindler in *The Algal Bowl:*

> *Overfertilization of the World's Freshwaters and Estuaries*. In an interview, Schindler said, "We really need to treat a feedlot with 50,000 cattle like we treat a small city." Pigs, cattle, and humans produce abundant waste that seeps into water systems, accumulating in lakes where it causes eutrophication: nutrient levels spike, causing algal blooms, which in turn deplete the water of oxygen, killing fish and other organisms. In Canada, this problem affects lakes located south of 51 degrees latitude in a band stretching from coast to coast.

You can also use a quotation from an unrelated source or author, as long as it is relevant to your topic and not so well known that it will appear trite. A dictionary of quotations can be helpful, allowing you to search for quotes by author, key words, or even topic.

The question

A rhetorical question will only annoy the reader if it's commonplace or if the answer is obvious, but a thought-provoking question can make a strong opening. For example, you might begin an essay on the issue of court fines for environmental damages with the question, "How do you put a price on the cultural and biological damage of chopping down the Golden Spruce on Haida Gwaii?" Just be sure that you do actually answer the question somewhere in your essay.

The anecdote or telling fact

This is the kind of concrete lead that journalists often use to grab their readers' attention. For example, a biology paper on collagen in scurvy patients might begin:

> By the time Magellan's ship, the *Trinidad*, had crossed the Pacific, only a dozen men were left functioning in his crew. Scurvy had taken 20 souls already. Another 20 were too sick from the disease to be of any use. Magellan was just one of many long-distance sailors whose crews would pay a high price for a diet too low in fresh fruits and vegetables.

Save this approach for your least formal essays—and remember that the incident must really highlight the ideas you will discuss in your essay.

Writing Conclusions

Endings can be painful. Too often, writers feel that they ought to say something profound and memorable, but end up writing a pretentious or affected ending, such as the following:

> In summary, gene therapy is going through a period of technical revolution. The advances being made in this field will undoubtedly save thousands of lives in the future.

Experienced editors say that many articles and essays would be better without their final paragraphs; in other words, when you have finished saying what you have to say, the best thing to do is to stop. This advice may work for short essays, where you can keep the central point firmly in the foreground. However, for longer pieces, where you have developed a number of ideas or a complex line of argument, you should provide a sense of closure. Readers welcome an ending that helps to tie the ideas together; they don't like to feel as though they've been left dangling. Because the final impression is often the most lasting, it's in your interest to finish strongly. Simply restating your thesis or summarizing what you have already said isn't forceful enough. The following sections discuss several alternatives.

The inverse funnel

The simplest conclusions restate the thesis in different words and then discuss its implications. Baker and Gamache call this the *inverse funnel approach*, as opposed to the *funnel approach* of the opening paragraph.[3] In this type of conclusion, the specific arguments made in the body of the essay widen to a more inclusive final statement.

The sample essay on phenotypic plasticity, which opened with a funnel approach, concludes with an inverse funnel:

> The evidence supplied in this essay shows that temperature, climate, water availability, day length, soil nutrition, insects, and diseases can all affect phenotypic plasticity. The complex gene interactions generated by these factors result in a wealth of developmental and adaptive responses. Far from a simple one gene–one phenotype model, organisms show a variety of reaction norms. This evidence suggests that phenotypic

plasticity is a key component of an organism's ability to adapt to its environment.

One danger in moving to a wider perspective is that you may try to embrace too much. When a conclusion expands too far, it tends to lose focus. It's always better to discuss specific implications than to trail off into vague generalities in an attempt to sound profound.

The new angle

A variation on the basic inverse funnel approach is to reintroduce your argument with a new twist. Suggesting a fresh angle can make your ending more compelling or provocative. Beware of introducing an entirely new idea, though, or one that's only loosely connected to your original argument; the result, if it's too far off-topic, could detract from your argument. The following example is a short and very effective conclusion to a highly focused essay on the function of tapetal cells:

> Ultimately, understanding the factors that control a tapetal cell's molecular and physical triggers will lead to deeper insights into conifer sterility. The tapetal layer is essential to conifer reproduction, as it creates the pollen wall. As the protective shell, the pollen wall protects the male gametes from both the elements and the surrounding environmental stresses. Although this structure has often been considered mere covering, its role is possibly more dynamic than previously thought.

The full circle

If you begin your essay by relating an anecdote, posing a rhetorical question, or citing a startling fact, you can complete the circle by referring to it again in your conclusion, relating it to some of the insights revealed in the main body of your essay. This technique provides a nice sense of closure for the reader.

The stylistic flourish

Some of the most successful conclusions end on a strong stylistic note. Try varying the sentence structure: if most of your sentences are long and complex, make the last one short and punchy, or vice versa. Sometimes you can dramatize your idea with a striking phrase or a colourful image. When you are writing your essay, keep your eyes open and your ears tuned for fresh ways of putting things, and save the best for the end. If you want good examples of short and long essays that finish with a flourish, have a look at articles published in the generalist journal *New Scientist*.

Summary or Conclusion—What's the Difference?

A summary restates the aim and then outlines the major premises of an essay. A conclusion may contain elements of a summary, but it often introduces opinion or perspective into the writing. Conclusions can discuss future research directions, or they can evaluate the merit of research in a particular field. Remember: you draw a conclusion, but you don't draw a summary. The former involves opinion, while the latter is a mere shopping list of highlights. Life sciences students find it much more difficult to write a conclusion than do students in the humanities; science students often feel that they cannot diverge from the objectivity and neutrality of scientific method.

So when should you end your essay with a conclusion rather than a summary? This decision will depend on your topic. When science methodology meets its limitations, when science meets or even creates controversy, or even when science meets society—and they tend to meet very often—then opinion and conjecture have their place. You might say that to not have an opinion in a conclusion leads the reader to conclude that the writer is either naive or uncritical.

Integrating Quotations

A quotation can benefit an essay in two ways. First, it can add depth and credibility by showing that your position or idea has the support of an authority. Second, it can provide stylistic variety and interest, especially if the quotation is colourful or eloquent. The trick to using quotations effectively is to make sure that you properly integrate them into your own discussion—they should neither dominate your ideas nor seem tacked on. To ensure that a quotation has the desired effect, always refer to the point you want the reader to take from it; don't let a quotation dangle on its own. Usually, the best way to do this is to make the point before the quotation:

> Shortly after receiving the Nobel Peace Prize in 1962, Linus Pauling expressed his dismay at the effects of radioactive fallout from nuclear bombs: "I like people. I like animals, too—whales and quail, dinosaurs

and dodos. But I like human beings especially, and I am unhappy that the pool of human germplasm, which determines the nature of the human race, is deteriorating."

Sometimes, however, a quotation can precede the explanation. This format is common in introductory paragraphs, as illustrated earlier in this chapter. You must make sure, however, that you explain the significance of the quotation so that your reader is not left wondering why you have added it.

If a complete sentence precedes the quotation, use a colon at the end of the introductory phrase; if you work the quotation into the syntax of your own sentence, you do not need to add any punctuation before the quoted material; to otherwise introduce a quotation, use a comma:

> John Polanyi put it best: "Nothing is more irredeemably irrelevant than bad science."

(or)

> Was John Polanyi being merely rhetorical when he said that "[n]othing is more irredeemably irrelevant than bad science"?

(or)

> John Polanyi felt strongly, insisting, "Nothing is more irredeemably irrelevant than bad science."

If the quoted passage comes at the end of a sentence, finish with a period before the closing quotation mark—unless the sentence is a question, in which case finish with a question mark after the quotation marks. If you need to adjust any element of the quoted material (e.g., changing a capital letter to a lower-case letter), place your modification in square brackets.

The length of a quotation can vary from a short phrase woven into the middle of a sentence to a paragraph or more. Just remember that the longer the quotation, the greater the danger that it will overshadow, rather than reinforce, your own viewpoint. Don't quote any more than you really need. You should also know that writers in certain disciplines, especially those in the life sciences, tend not to include direct quotations; when they do use them, the quotations are usually very short. If you're unsure about the conventions for your own discipline, you can always ask your instructor.

Academic Honesty and Avoiding Plagiarism

If you assemble your facts from peer-reviewed journals, acknowledge your sources, and do not claim more in your interpretations than is rightfully yours, then you will not face problems with academic honesty. However, it is possible to make errors. Although many of these errors can be subtle, it is best to clear up the more obvious challenges to academic honesty that can arise early in your studies.

Academic dishonesty has many egregious forms, including plagiarism, fraudulence, impersonation, sabotage, fabrication, and cheating, as well as some more subtle forms of deception and misrepresentation. Most types of academic dishonesty are simple and obvious, but dishonesty can also be complex. Although most dishonest acts provide a selfish reward, some are altruistic. An example is students in large classes with multiple sittings of exams, who provide test answers to friends in other sections. Altruism, in this case, erodes the value of education. Although dishonest lab practices can be conducted with malice, such as sabotaging someone else's experiment, others can be altruistic, such as "dry-labbing." In this case, students know that an instructor or teaching assistant is expecting a particular outcome. Students help one another reverse-engineer their lab results so that they meet the expectations. Alternatively, they never carry out the experiment at all; instead, they calculate data that would provide the correct outcome. This results in an optimized collective outcome for a team of lab partners. In this case, the false description, no matter how clever or altruistic, constitutes fabrication.

Professors, TAs, and instructors have little time for academic dishonesty. Science involves the pursuit of truth. Before a scientific manuscript can be published, it is first anonymously peer-reviewed. This allows uninhibited criticism. If the paper represents honest, original, well-executed work, it will be published. This filter on information keeps most scientific fields quite clean. In contrast, dishonest publication practices corrupt science. If left unchecked, corruption damages the reputations of scientists, which, in practical terms, results in an erosion of public trust and, eventually, public financial support. Reputations built slowly—paper by paper—are ruined when evidence of corrupt practices are discovered and proven. Life sciences students are young scientists trying to build solid reputations essay by essay, lab report by lab report. Students must be equally careful to think carefully about their ethical and moral standards and to not let the edifice of their accomplishments crumble.

Plagiarism is usually described as unattributed use of someone else's ideas or work. Plagiarism ranges from stealing someone else's important ideas, to failing to adequately state in one's own words what another author has written. Consequently, plagiarism has many grotesque forms, but also some more subtle forms. Grotesque plagiarism involves passing off the work of others as one's own: this includes such examples as purchased essays, copied assignments, and other forms of imposture. There is a more subtle level. Passing off someone else's ideas, even when written entirely in the student's own hand, is equally intolerable. In its simplest form, plagiarism means borrowing someone else's interpretation and not attributing these ideas to the original author or thinker.

Plagiarism flourishes in ambiguous situations, often because students do not default to the norm of science, which is to never use a piece of knowledge without attribution. If you use someone else's idea, acknowledge it, even if you have changed the wording or just summarized the main points. In the life sciences, your work will be more convincing if you acknowledge the ideas of others.

Let's say you are writing an essay on new directions in the field of ecology. The following passage comes from Charles Hall's article on ecology[4] in the *Encyclopedia of Earth*, a Creative Commons (cultural free-share) site:

> More recently, ecology has included increasingly the human-dominated world of agriculture, grazing lands for domestic animals, cities, and even industrial parks. Industrial ecology is a discipline that has recently been developed, especially in Europe, where the objective is to follow the energy and material use throughout the process of, e.g., making an automobile with the objective of attempting to improve the material and energy efficiency of manufacturing.

One student's essay includes the following passage. It is plagiarized because the student repeated exact phrases from the original article without acknowledging the source:

> ✗ Ecology has begun to include **the human-dominated world of agriculture, grazing lands for domestic animals, cities, and even industrial parks**. Additionally, **industrial ecology** as **a discipline, especially**

in Europe, follows **energy use throughout the process of** tproduction, such **as making an automobile. The objective is to attempt to improve the material and energy efficiency of manufacturing.**

To avoid a charge of plagiarism and its unpleasant and sometimes disastrous consequences, all you need to do is acknowledge your source. In the correctly documented passage below, the writer has used quotation marks to identify words and phrases taken directly from the original article; the writer has also included a parenthetical text citation. (See Chapter 11 for alternative citation styles.) In this case, the writer would provide a bibliography at the end of the essay that gives complete publication information for the source.

✔ According to Hall (2009), ecology has begun to include the "**human-dominated world of agriculture, grazing lands for domestic animals, cities, and even industrial parks. Industrial ecology is a discipline that has recently been developed, especially in Europe, where the objective is to follow the energy and material use throughout the process of, e.g., making an automobile with the objective of attempting to improve the material and energy efficiency of manufacturing.**"

In the life sciences, unlike in the arts and humanities, writers usually avoid using direct quotations. As a student, you must learn to interpret what you read and rewrite it in your own words, a process that often involves reorganizing and synthesizing information from a number of different sources. (For examples of proper paraphrasing and a name–year style of in-text citation in a life sciences essay, see the sample annotated student essay on pages 68-87.)

In the following passage, this student has made the mistake of assuming that putting the information in his or her own words is good enough. It's not. The concept of industrial ecology is still "borrowed":

✘ Ecology has shifted some of its focus to agriculture and industry. Industrial ecology is a recent development that uses ecological models to study the energy efficiency of manufacture.

Remember that plagiarism involves not only using someone else's words but also expressing their ideas—even when using your own words—without referencing the original source.

In the correctly documented passage below, proper acknowledgement takes the form of an in-text reference to the original author:

- ✔ Recently, ecology has begun to include studies of human activities, particularly agriculture and industry. An example is the new field of industrial ecology that creates models of the energy efficiency of manufacturing processes (Hall 2009).

Where should you draw the line on acknowledgements? As a rule, you don't need to give credit for anything that's common knowledge. You wouldn't provide a reference for the fact that DNA is present in a cell's nucleus or that Darwin and Wallace developed theories of evolution, but you should acknowledge any idea that is neither well known nor your own. Always document any fact or claim—statistical or otherwise—that is unfamiliar or open to question.

For all students, online material presents a particular hazard and can cause a lot of grief. Even though websites are instantly accessible, the material is not common property. In fact, online material is the property of the individual or organization that publishes it and is often protected by copyright in the same way that printed material is. Entries on websites can be cited like any other record. (Information about the proper procedure for documenting online material is included in Chapter 11.) Even if material is free, such as on Wikipedia, it is crucial that you properly acknowledge the information you use.

More commonly, websites are mined for illustrations—for example, Wikimedia Commons provides a large collection of images and other media files. You *must* acknowledge the source of all illustrations that you did not draw, photograph, or otherwise personally produce. In all scientific papers published in trade journals, the illustrations are done by the author(s). If not, the figure caption provides an acknowledgement, which usually states that permission was granted by the journal in which the illustration was first published. You may be able to use photographs or illustrations from other sources, but clear it first with your instructor or TA.

A last note on plagiarism: in the event you are falsely accused, remember that the best defense is a good offense. Counter any accusation with evidence: hand in either a file folder containing your research notes or a USB key with old drafts. This should clear things up immediately.

The Editing Stage

Often the best writer in a class is not the one who can dash off a fluent first draft but the one who is the best editor. To edit your work well you need to see it as the reader will, and in order to do that you have to distinguish between what you meant to say and what you have actually put on the page. For this reason it's a good idea to leave some time between drafts so that when you begin to edit you will be looking at the writing afresh rather than reviewing it from memory. This is the time to do something that will take your mind off your work, to go to a movie or the gym. Without this distancing period, you can become so involved in your paper that it's hard to see your writing objectively.

Editing doesn't mean simply checking your work for errors in grammar or spelling. It means looking at the piece as a whole to see if the ideas are well organized, well documented, and well expressed. It may mean making changes to the structure of your essay by adding some paragraphs or sentences, deleting others, and moving others around. Experienced writers may be able to check several aspects of their work at the same time, but if you are inexperienced or in doubt about your writing, it's best to look at the organization of the ideas before you tackle sentence structure, diction, style, and documentation.

What follows is a checklist of questions to ask yourself as you begin editing. Far from all-inclusive, it focuses on the first step: examining the organization of your work. Because you probably won't want to check through your work separately for each question, you can group some together and overlook others, depending on your own strengths and weaknesses as a writer.

Checking for organization

- Is the title concise and informative?
- Are the purpose and approach of this essay evident from the beginning?
- Are all sections of the paper relevant to the topic?
- Is the organization logical?
- Are the ideas sufficiently developed? Is there enough evidence, explanation, and illustration?
- Would an educated person who hasn't read the primary material understand everything? Do any points need additional clarifications or explanations?
- Does the discussion take into account opposing arguments or evidence?

- Do the paragraph divisions make the ideas more coherent? Do these divisions signal where one idea transitions to the next? Are similar ideas grouped together in each paragraph?
- Do any parts of the essay seem disjointed? Would more transitional words or logical indicators make the sequence of ideas easier to follow?
- Do the conclusions accurately reflect the argument in the body of the work?

If you have difficulty visualizing the overall organization of your paper on-screen, try printing the pages and then placing them beside one another on the floor, bed, or desk. This process should help you to see what is not always apparent on a computer—namely, the number of lines or pages that you have devoted to various parts of your argument or description. You can then decide whether any of your discussions seem too long or too short.

You can also devise your own checklist based on comments you have received on previous assignments. These questions will be particularly useful when you move from the overview of your paper to the close focus on sentence structure, diction, punctuation, spelling, and style. If you have a particular weak area—for example, irrelevant evidence or run-on sentences—you should give it special attention. Keeping a personal checklist will save you from repeating the same old mistakes.

Your word-processing program will catch typing as well as spelling errors, but remember that it may not point out actual words that are used incorrectly (such as *borne* when you need *born*). Think of a computer-based spell-check as a useful first pass rather than a final one. Similarly, grammar checkers are not reliable. They will pick up common grammar errors and stylistic problems, but they do make mistakes and will likely never equal the judgment of a good human editor.

Always read over your work with a critical eye and take the time to change anything that is unsatisfactory. If you have doubts about your essay, read it aloud to yourself—boring bits, illogical lines of argument, poor style, awkward phrasing, and bad grammar will become immediately obvious. If you're still unsure, try recording your reading and replaying it to yourself, or read your essay to a friend, loyal roommate, or willing family member.

Keep in mind, too, that for final editing most good writers suggest working from the printed page rather than from the computer screen. Print a draft and use the hard copy for your editing and proofreading. You will read more slowly and with greater acuity, and the final product will be more polished as a result.

Formatting Your Essay

It is important to ensure that the visual presentation of your essay is clear and professional. Consider the following guidelines when formatting your paper:

- **Use a consistent page layout.** Number each page. Always double-space your work and use margins of at least one inch to frame the text in white space and allow room for your reader to write comments.
- **Create a cover page.** Provide a neat, well-spaced cover page that includes the title, your name, and the name of your instructor and course.
- **Select appropriate fonts.** Make good use of formatting features such as bold and italics for emphasis, and choose appropriate font faces. One rule of thumb is to use a serif font such as Times New Roman, usually 12-point, for the body of the essay and a sans-serif font such as Arial for the title and headings. However, most instructors will provide you with guidelines specifying their particular preferences for presentation and formatting.

Sample Annotated Student Essay

The following essay on restoration of dam sites using alder (pages 68–87) illustrates various style problems that are common in student-written essays. As you read through the essay, examine the annotated comments and think about how they might help you to improve your own work. Compare the original "bad" version with the revised "good" version—you will notice what an improvement a few simple edits can make. The student received a B for the original essay—the content was good, but the writing contained many problems with style, grammar, and overall effect. After receiving her instructor's suggestions, the student re-edited and then resubmitted the paper. The "good" version received an A. In almost every case, improving your style and grammar will earn you a higher grade. Remember that you won't often have the opportunity to resubmit your essay for a higher grade, as this student did, so edit your work as carefully as you can before handing it in.

In scientific writing, starting a sentence with a number or an abbreviation is not generally acceptable.

BAD

Use of abbreviations is inconsistent throughout the essay. Once an abbreviation is established, it should be used.

Alder used in restoration of landscapes in Northern Quebec

1971 marked the beginning of unprecedented hydroelectric project by the Government of Quebec. La Grande Complex, also known as the James Bay Project, is fed by the La Grande River basin, which has a total area of 177,000 km² across northern Quebec. To this day, it remains one of the largest hydroelectric projects in the world. Such a grand undertaking was not without impact on the environment. Redirection of some rivers and flow reduction of others left riverbanks barren and prone to erosion and land clearing for roads and construction projects affected expanses of boreal forests. The history of the project's development has been characterized by conflict between government agencies and First Nations people in Northern Quebec due to disruption to their fishing and hunting activities (Hornig, 1999). In 1976, the first of the land reclamation projects began that aimed to re-vegetate thousands of hectares of land that had been cleared for construction of roads and dams, as well as planting in river beds that were now exposed due to reduced water flow. The harsh, cold climate of the Canadian North made this not an easy task, but one solution was beginning to surface in the mid 70's: actinorhizal symbioses, a complex relationship between a bacterium and accommodating trees, such as green alder. This new phytotechnology allowed the Société d'énergie de la Baie James (SEBJ), a Hydro-Québec subsidiary tasked with construction of the dam complex, to replant thousands of hectares of boreal woodlands with green alder (Hayeur, 2001). This paper will describe actinorhizal plants and their role in reclamation of lands in the James Bay Project.

This paragraph suffers from having too many long sentences in a row.

GOOD

Alder used in restoration of landscapes in Northern Quebec

In 1971, the Government of Quebec launched an era of unprecedented hydroelectric project construction. La Grande Complex, also known as the James Bay Project, is fed by the La Grande River basin, which covers an area of 177,000 km² across northern Quebec. To this day, the James Bay Project remains one of the largest hydroelectric projects in the world. Such a grand undertaking was not without impact on the environment. Redirection of some rivers and flow reduction of others left riverbanks barren and prone to erosion. Land clearing <u>affected large expanses</u> of boreal forests, disrupting fishing and hunting activities of First Nations people in Northern Quebec. <u>Land reclamation was a solution to one of the many conflicts that arose</u> between government agencies and First Nations people (Hornig, 1999). In 1976, the first of the land reclamation projects began. The goal was to re-vegetate thousands of hectares of land and <u>to vegetate the now exposed river beds</u>. The harsh, cold climate of the Canadian North <u>posed a challenge</u>. One solution that surfaced in the mid 70's was to exploit actinorhizal symbiosis, which is the complex relationship between <u>a bacterium, *Frankia*, and its host tree, green alder</u>. This new phytotechnology allowed the Société d'énergie de la Baie James (SEBJ), a Hydro-Québec subsidiary tasked with construction of the dam complex, to replant <u>former woodlands and riverbeds</u> with green alder (Hayeur, 2001). This paper will describe actinorhizal plants and their role in reclamation of lands in the James Bay Project.

1

Generally, run-on sentences have been broken up. The underlined material denotes instances where the author has corrected inaccuracies.

BAD

Re-vegetation following disturbances in Quebec's northern climate is slow-going. Trees in this territory are generally small, reaching 10–15 cm in diameter and are not generally prized by commercial forestry. Winter temperatures average around –23°C and can reach –50°C. Summers are warm with averages of around 14°C, reaching a maximum of around 34°C. The geological history of the region has been shaped by glacial activity, which left hilly moraines and gravel deposits in depressions. These unforgiving conditions presented the Société d'énergie de la Baie James with challenges (Hayeur, 2001).

The objective of this paragraph is unclear, and the first line needs to be rewritten.

GOOD

Re-vegetation of disturbed sites is a slow process in the challenging environments of northern Quebec. Trees are not generally prized by commercial forestry, because they tend to be small. Summers are warm with averages of around 14 °C, reaching a maximum of around 34 °C. However, winter temperatures average around –23 °C and can reach –50 °C. The geological history of the region has been shaped by glacial activity, which resulted in extensive deposits of rocks and gravel. The landscape is one of gravel depressions and hilly moraines. Soils are young, thin and nitrogen-poor. These unforgiving climate and soil conditions presented the SEBJ with numerous biological problems (Hayeur, 2001).

The first sentence of the paragraph has greater clarity and is, therefore, more powerful than its predecessor. Sentences were restructured and relocated to create a better narrative.

BAD

Paragraph beginnings are generally weak.

A problem in many essays is assuming the reader will make all the connections.

Around the time the James Bay Project was beginning, the field of land reclamation had become interested in the use of actinorhizal plants for revegetation of land with poor soil (Normand, 2013). Actinorhizal plants are form root nodules in a symbiotic relationship with *Frankia*, a genus of gram-positive nitrogen-fixing bacteria. The plant benefits from nitrogen made available by the bacteria and the bacteria makes use of photosynthates manufactured by the plant. *Frankia* enters its host plant either through root hairs or through intracellular infection. The infection causes the root cortex and hypodermis to divide and form a pre-nodule which allows hyphae from *Frankia* strains to enter the root cortex. Colonizing hyphae enter the plant's cell walls and proliferating hyphae branch out into the host plant's cytoplasm. A root nodule forms at the infection site. *Frankia* hyphae give rise to spherical, terminal, or lateral vesicles that provide the site of nitrogenase activity for nitrogen fixation (Benson et al., 1993).

Final sentences in a paragraph complete an idea developed in the paragraph, but they may also serve as a bridge to drive the storyline. A sentence is needed here to keep the momentum.

GOOD *To strengthen the narrative, this portion was fleshed out. The critically important definition was linked to the central thesis of the essay.*

Just around the time the James Bay Project began, a potential solution to the soil nutrition problem was found: actinorhizal plants (Normand, 2013). These plants formed root nodules with *Frankia*, a genus of gram-positive nitrogen-fixing bacteria, creating a mutually beneficial relationship, i.e., a symbiosis. The symbiosis depends on life forms being able to use molecules from the atmosphere and not the soil. Bacteria fix nitrogen from the air, making it available to the plant in an easily transportable organic compound. In exchange the plant provides the bacteria with sugars manufactured from atmospheric carbon dioxide. *Frankia* enters its host plant either through root hairs or through intracellular infection. The infection causes the root cortex and hypodermis to divide to form a pre-nodule, which accommodates more hyphae of *Frankia* entering the enlarging root cortex. Colonizing hyphae enter the plant's cell walls and proliferating hyphae branch out into the host plant's cytoplasm. A root nodule forms at the infection site. *Frankia* hyphae give rise to spherical, terminal, or lateral vesicles that are the sites of nitrogen fixation (Benson et al., 1993). Initially, creating the symbiosis at a commercial scale was challenging. Host plants were easy to grow, but bacteria proved problematic.

A bridging sentence was created.

BAD

Prior to 1978, hundreds of attempts to isolate *Frankia* strains were met with difficulties due to their slow growth rate and the uncertainty about appropriate growing medium. Success was finally achieved when the strain CpI1 was isolated from root nodules of sweetfern, *Comptonia peregrina* (Callaham et al., 1978). This strain successfully re-infected its host. A further advance was made when it was discovered that *Frankia* strains present in root nodules of green alder (*Alnus crispa*) were able to infect other green alder seedlings. In the end, SEBJ chose this species, because of the relative ease with which it could be infected (Benson et al., 1993).

Paragraph beginnings are weak. Note how many paragraph beginnings in this essay start with a short clause. It is stylistically weak, because it is used repetitively throughout.

GOOD

Hundreds of attempts were needed to isolate *Frankia* strains. The bacteria grew very slowly. The conditions leading to symbiosis were very strain-dependent. Success was finally achieved when the strain CpII was isolated from root nodules of sweetfern, *Comptonia peregrina* (Callaham et al., 1978). This strain successfully re-infected its host. A further advance was made when it was discovered that *Frankia* strains present in root nodules of green alder (*Alnus crispa*) were able to infect other green alder seedlings. In the end, SEBJ chose this species, because of the relative ease with which it could be infected (Benson et al., 1993).

BAD

SEBJ was an early pioneer in the large-scale use of actinorhizal plants for reclamation technology. In 1973, before the symbiosis was understood, green alder had been planted in a successful reclamation trial of lands surrounding the Manic-5 dam completed. The project was a joint effort between Hydro Quebec, SEBJ's parent company, and researchers at Laval University. This initial success made possible further research and development (Périnet et al., 1985). As a result, SEBJ became an early pioneer in the large-scale use of actinorhizal plants for reclamation technology (Hayeur, 2001).

The objective of this paragraph is unclear in terms of the rest of the paragraph. The first line needs to be rewritten or relocated.

GOOD

The choice of green alder was not entirely fortuitous. In 1973, before the symbiosis was understood, green alder had been planted in a successful reclamation trial of lands surrounding the Manic-5 dam completed. The project was a joint effort between Hydro Quebec, SEBJ's parent company, and researchers at Laval University. This initial success made possible further research and development (Périnet et al., 1985). As a result, SEBJ became an early pioneer in the large-scale use of actinorhizal plants for reclamation technology (Hayeur, 2001).

5

The first sentence was moved to the end and replaced with a better opening sentence.

BAD

Dates are distracting, forcing the reader back and forth.

Collaboration with Laval University and a spin-off company, Rhizotec Laboratories, allowed Hydro-Quebec to successfully demonstrate that actinorhizal alder seedlings could be produced on an industrial scale (Hayeur, 2001). Before 1980, the norm was to crush and homogenize field collected nodules in a buffer to manufacture inoculum. In 1980, however, isolated *Frankia* strains became commercially available and Laval University was able to prepare pure *Frankia* culture inoculum for SEBJ, releasing the potential for contamination by unwanted microorganisms and pathogens. These experiments also established the practice of using overhead sprayers to inoculate seedling on a large scale. This was a fantastic breakthrough for SEBJ, allowing them to produce nodulated seedlings on a massive scale. Between the years of 1979 and 1984, seven million inoculated seedlings were produced. Six million of these were planted during the La Grande project and survival for the seedlings was 90–95% (Périnet, 1985).

This paragraph suffers from having too many long sentences in a row.

GOOD

Collaboration with Laval University and a spin-off company, Rhizotec Laboratories, allowed Hydro-Quebec to demonstrate that actinorhizal alder seedlings could be successfully produced (Hayeur, 2001). Initially, inoculum was made from field-collected nodules that had been homogenized in a buffer. These crudely extracted inocula were often contaminated with microorganisms and pathogens. After 1980, these crude mixtures were replaced by pure strains grown in laboratory conditions. A further advance was the use of overhead sprayers to inoculate thousands of seedlings simultaneously. Of the six million seedlings planted during the La Grande project, approximately 95% survived (Périnet, 1985).

6 *The distraction of unnecessary dates was eliminated, which allowed the contrast between crude and pure extracts to be brought into focus.*

BAD

Poplars do not need to be mentioned as the paragraph is all about alders.

Since the early 80s, actinorhizal plants have been investigated for a number of other industrial scale projects including revegetation of mine sites, improvement of agroforestry systems, stabilization of dunes, establishment of windbreaks and reclamation of oil sands (Diagne et al., 2013). In 1990, poplars were successfully used in strip mine reclamation thanks to their high water uptake and ability to quickly reintroduce large amounts of organic matter into the soil, both of which result in reduced contamination migration (Wheeler et al., 1990). Problems have arisen, however, as alders are known to acidify soil which can increase the bioavailability of contaminants such as copper. This effect could be helped with the addition of liming (Fessenden and Sutherland, 1979). In the right conditions and when combined with the right companion plants, however, alders can be used as nurse plants and acidification can be helpful to increase heavy metal phytoextraction by other vegetation. Alders, coupled with some strains of *Frankia*, have been found to benefit from presence of nickel in soil as they require a large amount of nickel for nitrogen fixation activities. Others are extremely sensitive to nickel contamination and exhibited significant reduction of nodule mass and nitrogen fixation rates (Roy et al., 2007). Actinorhizal alders have also been shown to improve soils. Soil microbial diversity increased. Planting alder associated with *Frankia*

There are problems of attribution. It is unclear whether "Others" refers to alders or to activities.

BAD

This sentence is inappropriately placed and should not end this paragraph, as it breaks the flow of ideas from this paragraph to the next.

was an effective method of soil improvement in Alberta oil sands (Lefrançois et al., 2010) and in gold mine tailings (Callender et al., 2016). Genomic techniques were used to identify appropriate plant-*Frankia* combinations. Going forward, the effects of climate change on alders are being examined to better prepare for changing environments (Tobita et al. 2016).

This paragraph can be easily incorporated in the previous one.

Coupled with the wide variety in *Frankia* genetics and physiology, is a high variability in tolerance to different environmental stresses, such as pH, temperature, and the presence of heavy metals. Most *Frankia* isolates show optimum growth between pH of 6 and 8, however, strain BMG5.1 show optimal growth at pH 9.5 and CpI1 can grow at a pH of 4.6 (Ngom et al., 2016). This illustrates the need for careful selection of host and bacterial symbionts based on specific conditions of the target site (Roy et al., 2007).

The writer provides actual data derived from two Frankia isolates and cites Ngom et al., 2016, as the source. Because the writer was required to cite mostly primary sources, i.e., actual studies, passing off Ngom et al.—a review—as a study is the kind of carelessness that suggests inaccuracy.

GOOD

Actinorhizal plants have been investigated in a number of other industrial-scale projects including re-vegetation of mine sites, improvement of agroforestry systems, stabilization of dunes, establishment of windbreaks and reclamation of oil sands (Diagne et al., 2013). Problems have arisen, however, as alders caused soil acidification, which increased the bioavailability of contaminants such as copper. This effect could be reduced by the addition of lime (Fessenden and Sutherland, 1979). In the right conditions and when combined with the right companion plants, alders have been used as nurse plants for other plants able to extract heavy metals. Metals may even be beneficial. Because a significant amount of nickel is required for nitrogen fixation activities, alder-*Frankia* association functioned better in the presence of this metal. However, this effect was not universal, as some alder-*Frankia* combinations were extremely sensitive to nickel, resulting in a significant reduction of nodule mass and nitrogen fixation rates (Roy et al., 2007). Actinorhizal alders have also been shown to improve soils. Soil microbial diversity increased. Planting alder associated with *Frankia* was an effective method of soil improvement in Alberta oil sands (Lefrançois et al., 2010) and in gold mine tailings (Callender et al., 2016). *Frankia*'s ability to tolerate environmental stresses, including pH

7

GOOD

changes, temperature fluctuations, and metals in soils is due to high genetic variation as well as physiology. A few examples show how variable these responses can be. Although most *Frankia* isolates show optimum growth between pH of 6 and 8, one strain, BMG5.1, shows optimal growth at a high pH of 9.5 and another strain, CpI1, can grow at a low pH of 4.6 (see review by Ngom et al., 2016). This illustrates the need for careful selection of host and bacterial symbionts based on specific conditions of the target site (Roy et al., 2007). Modern genomic techniques will undoubtedly contribute to the identification of appropriate plant-*Frankia* combinations. From the plant side, effects of climate change on alders are being examined to better prepare for changing environments (Tobita et al., 2016). As our knowledge concerning the physiology and genetics of this association improves, we will be better positioned to make optimal selections of combinations of alder and bacteria.

Combining the paragraphs provides greater unity of thought.

The writer has changed the citation from "Ngom et al., 2016" to "see review by Ngom et al., 2016," thereby identifying it as a secondary source and dispelling confusion. An even better solution would be to read and then cite the primary sources for the data, namely Gtari et al., 2015, and Burggraaf and Shipton, 1982.

BAD

Actinorhizal plants provide an advantage in the reclamation of degraded and contaminated lands, because they do not require use of expensive fertilizers: nitrogen in the air is free to a *Frankia* bacterium in its alder home. Increasingly, re-vegetation strategies involving actinorhizal plants are becoming attractive. However, assessment of the long-term effects of such projects is still lacking. For example, the SEJB have provided no information—at least in English—about the eventual state of their reclamation efforts. More research and follow up on current projects is needed to identify appropriate combinations for tackling specific environmental conditions.

The last paragraph provides a new global perspective, but it ends weakly. The writer is struggling to square future applications with the goals of the essay. It would have been better if the writer had revised the closing paragraph to link with the objective of the essay. Sadly, the ending limps to a conclusion.

GOOD

Actinorhizal plants provide an advantage in the reclamation of degraded and contaminated lands, because they do not require use of expensive fertilizers: nitrogen in the air is free to a *Frankia* bacterium in its alder home. Increasingly, re-vegetation strategies involving actinorhizal plants are becoming attractive. However, assessment of the long-term effects of such projects is still lacking. For example, the SEJB have provided no information—at least in English—about the eventual state of their reclamation efforts. More research and follow up on current projects is needed to identify appropriate combinations for tackling specific environmental conditions. Even so, the clear benefits of using nitrogen-fixing bacteria associated with alder nodules is a reminder that complex interspecies relationships—once understood—can provide humanity with new biologically based, environmentally friendly opportunities to repair the damage that accompanies the ever-expanding population of humans.

9

The doubt is created, which is answered. Adding a moment of tension in a concluding paragraph is a much easier way to pave the way for confirming the thesis statement at the beginning of the essay.

GOOD

References

Benson, D.R., Silvester, W.B. 1993. Biology of *Frankia* strains, actinomycete symbionts of actinorhizal plants. Microbiol Reviews 57: 293–319.

Callaham, D., Del Tredici, P., Torrey, J. G. 1978. Isolation and cultivation in vitro of the actinomycete causing root nodulation in *Comptonia*. Science 199: 899–902.

Callender, K.L., Roy, S., Khasa, D.P., Whyte, L.G., Greer, C.W. 2016. Actinorhizal alder phytostabilization alters microbial community dynamics in gold mine waste rock from Northern Quebec: a greenhouse study. PLoS ONE 11: 0150181.

Diagne, N., Arumugam, K., Ngom, M., Nambiar-Veetil, M., Franche, C., Narayanan, K.K., Laplaze, L. 2013. Use of *Frankia* and actinorhizal plants for degraded lands reclamation. BioMed Research International 2013. http://dx.doi.org/10.1155/2013/948258

Fessenden, R.J., Sutherland, B.J. 1979. The effect of excess soil copper on the growth of black spruce and green alder seedlings. Botanical Gazette 140: S82–S87.

Hayeur, G. 2001. *Summary of knowledge acquired in Northern Environments from 1970 to 2000*. Hydro-Quebec Technical Report, Montreal, Quebec, Canada.

10 — *Omissions in reference sections can sometimes be indicators of plagiarism. This student, however, used reference management software such as EndNote or Mendelay. These types of software reduce errors substantially and students are encouraged to use them.*

GOOD

Hornig, J.F. 1999. *Social and environmental impacts of the James Bay Hydroelectric Project* (Vol. 18). McGill-Queen's Press-MQUP, Montreal, Quebec, Canada.

Lefrançois, E., Quoreshi, A., Khasa, D., Fung, M., Whyte, L.G., Roy, S., Greer, C.W. 2010. Field performance of alder-*Frankia* symbionts for the reclamation of oil sands sites. Applied Soil Ecology 46: 183–191.

Ngom, M., Oshone, R., Diagne, N., Cissoko, M., Svistoonoff, S., Tisa, L.S., Laplaze, L., Sy, M.O., Champion, A. 2016. Tolerance to environmental stress by the nitrogen-fixing actinobacterium *Frankia* and its role in actinorhizal plants adaptation. Symbiosis 70: 17–29.

Normand, P. 2013. A brief history of *Frankia* and actinorhizal plants meetings. Biosciences 38: 677–684.

Périnet, P., Brouillette, J.G., Fortin, J.A., Lalonde, M. 1985. Large scale inoculation of actinorhizal plants with *Frankia*. Plant and Soil 87: 175–183.

Roy, S., Khasa, D.P., Greer, C.W. 2007. Combining alders, *Frankia*e, and mycorrhizae for the revegetation and remediation of contaminated ecosystems. Botany 85: 237–251.

Tobita, H., Yazaki, K., Harayama, H., Kitao, M. 2016. Responses of symbiotic N_2 fixation in *Alnus* species to the projected elevated CO_2 environment. Trees: Structure and Function 30: 523–537.

11 As you can see, the revised essay is shorter. That's what a good edit will do—cut out the dross. Some students feel that they need to meet the word requirement in their first draft, but it's always better to be clear. If you find that your essay is too short after you've edited it, look for places where you could add useful content—for example, by adding the suggested paragraph about disease.

Protecting Your Work

Keeping backup copies of your work will protect you from losing it as a result of a computer or disk problem. The following practices will ensure that your work is adequately protected:

- Save regularly to protect against software or computer failure—once every 15 minutes is a good policy.
- Create a backup once a day, so that you have copies on both your hard drive and a USB flash drive or a server.
- Keep a copy of your file at least until you receive your grade for the course.
- Print an extra hard copy just to be on the safe side.

Getting into the habit of following these steps is a good investment of your time and will give you peace of mind.

Summary

Any piece of writing inevitably begins with an introduction and ends with a conclusion—pay attention to these sections because they create first and last impressions. Whenever possible, try to tie your introduction to your conclusion to create a unified essay. If you incorporate someone else's material, use quotations correctly and cite literature when appropriate to avoid the crime of plagiarism. When editing your work, pay particular attention to improperly cited material, which can lead to accusations of plagiarism. Make sure to review your organization to ensure that you present your ideas logically. Always leave enough time to polish your work so that you will receive the best mark possible—this is where style comes into play, as you have already taken care of the content. Remember to make backup copies of your assignments, both during the writing process and at the time of submission. It's always a good idea to keep your notes, as well as an extra hard copy and a digital copy of every assignment, until you get your final mark for the course.

Notes

1. John R. Trimble, *Writing with Style: Conversations on the Art of Writing* (Upper Saddle River, NJ: Prentice-Hall, 2000), 22–3.
2. Sheridan Baker and Lawrence B. Gamache, *The Canadian Practical Stylist*, 4th ed. (Don Mills, ON: Addison-Wesley, 1998), 55–6.
3. Baker and Gamache, 63–5.
4. Charles Hall, "Ecology," in *Encyclopedia of Earth* (http://www.eoearth.org/article/Ecology).

5 Writing a Lab Report

Objectives
- Assessing the purpose and the reader
- Understanding the format
- Learning the style, section by section
- Formulating a hypothesis
- Avoiding common pitfalls

Students in the life sciences write more lab reports than essays. Reports are based on the results of scientific experiments. Although lab reports generally conform to a basic format, every scientific discipline (chemistry, physics, biology, psychology, etc.) has slightly different requirements.

Any kind of academic writing should be clear, concise, and forceful, but for scientific writing there is one more imperative: be objective. Scientists are interested in exact information and the orderly presentation of factual evidence to support theories or hypotheses. When making a case for a particular hypothesis, you need to separate the facts you are reporting from your own speculation. You must never allow your preconceived opinions or expectations to interfere with the way you collect or present your data. If you do, you run the risk of distorting your results.

Always ask yourself the following questions: "Would anyone with an adequate background be able to repeat my results if they followed my instructions?" and "Would their interpretation be similar?" If you can answer "yes" to both questions, you have likely succeeded in remaining unbiased in your interpretations. You must conduct your experiment as objectively as possible and present the results in such a way that anyone who reads your report or attempts to duplicate your procedure will be likely to reach the same conclusions.

Purpose and Reader

You will most often write lab reports to demonstrate that you understand a theory or a phenomenon or that you know how to test a certain hypothesis. Because your reader is either an instructor or a teaching assistant, you can assume that he or she will be familiar with scientific terms; therefore, you do not need to define or explain terms if they are common knowledge in the field, such as *DNA, polymerase chain reaction*, or *natural selection*. Unusual terms should be defined—for example, *circumnutation* (i.e., helical stem movement during growth of plants). You should take care to use precise scientific terms— for example, rather than describing a lizard basking in the sun as a warm cold-blooded animal, describe it as an endothermic poikilothermic animal. While scientific terminology may be cumbersome, it is precise. You can also assume that the person marking your report will be on the lookout for mistakes in methodology or analysis and omissions of important data. Usually your reader will expect you to give details of your calculations, but even when you only need to provide the results of your calculations, you should be sure to note any irregularities in the experiment that might affect the accuracy of your results.

Format

Because the information in scientific reports must be easy for the reader to find, it should be organized into separate sections, each with a heading. One of the differences between writing lab reports and writing essays is that in reports you must use headings and subheadings as well as graphs, tables, or diagrams (see Chapter 10). By convention, most lab reports follow a standard order:

1. *Title Page*
2. *Abstract*
3. *Introduction*
4. *Materials and Methods*
5. *Results*
6. *Discussion*
7. *Conclusions*
8. *References*
9. *Attachments or Appendices*

The order of these sections is always the same, and it matches that found in journal articles. You may decide to combine sections (e.g., *Results and Discussion*), or you may want to separate sections (e.g., *Materials* presented separately from *Methods*); you might even choose to give the sections slightly different names, depending on what type of information you cover in each one. Different disciplines also have slightly different rules, but the following discussion will give you an overview of what you should include in each section of your report.

Title page

The title page is always the first page of the report. It should include your name, the title of the experiment, the date on which the experiment was performed, and the date of submission; for practical purposes, it should also include the name of your course and the name of your instructor. Your title should be brief—no more than 10 or 12 words—but informative, and it should clearly describe the topic and scope of your experiment. Avoid meaningless phrases, such as "A Study of . . ." or "Observations on. . . ." Simply state what it is you are studying, for example, "Enzyme Kinetics of 6-Phosphofructo-1-Kinase." Sometimes you may want to emphasize the result you obtained, for example, "Specialized Techniques for Mark–Recapture of Mammals, Especially Rabbits, in Winter."

Abstract

Your instructors might not ask you to write an *Abstract* section for most lab reports, but when you finally do have to create one, remember that although this is one of the first elements to appear in the lab report, it pays to write it last. An abstract is easier to write after you have completed all of the other sections because you know more precisely what you are reporting. If the abstract is written first, it will, inevitably, have to be revised substantially.

The abstract appears on a separate page following the title page. It is a brief but comprehensive summary of your report that should be able to stand alone; that is, someone should be able to read it and know exactly what the experiment was about as well as what the most important results were. The abstract should never mention any literature and usually does not contain any interpretation or discussion. Your summary should describe the purpose of the experiment, the experimental materials, the procedure, the results, and your conclusions. For a simple experiment, your abstract may be only a few lines, but even for a complex one you should keep it to less than 200 words. For this reason, you should avoid vague or wordy phrases—for example, don't say, "The reason for conducting the experiments in this study of X was to examine the effect of . . ." when you can be more concise: "X was studied to examine the effect of. . . ."

You must always write abstracts in the third person and in the past tense, and you can often include passive constructions that emphasize the subject of the experiment rather than the researcher. Remember: brevity is the goal of this section. To achieve this goal, you will need to draft and redraft your writing. Consult journal abstracts to get a feel for the impersonal, tight writing style required. Imitation is an excellent way to learn this somewhat awkward compositional style.

Here is an example of an abstract that summarizes an experiment involving proteomics to identify the proteins that might be found in gymnosperm pollination drops. The abstract introduces the experiment, describes the methods and materials, and summarizes the key results and conclusions:

> Pollination drops are a formative component in gymnosperm pollen–ovule interactions. Proteomics offers a direct method for protein discovery during this early stage of sexual reproduction. Pollination drops were collected from *Pinus sylvestris* L. by micropipette using techniques focused on preventing sample contamination. Drop proteins were separated from one another using both gel (SDS PAGE) and gel-free methods (liquid–liquid extraction, HPLC, capillary LC). Tandem mass spectrometric methods were used, including: LC/ESI-MS/MS on a QqTOF, a triple quadrupole, and an Orbitrap. Protein identification tools included Mascot, PEAKS, Analyst, and MS BLAST. Gel separation and MS/MS analysis of drops resulted in the identification of five proteins: thaumatin-like protein, serine carboxypeptidase, chitinase, xylosidase and defensin.

To avoid common pitfalls,

- make sure that your *Abstract* doesn't read like an *Introduction* section;
- do not include implications for future work; and
- only mention material that appears in the report.

Introduction

The *Introduction* section gives a more detailed statement of purpose or objective. It should describe the problem you are studying, the reasons for studying it, and the research strategy you used to obtain the relevant data.

If, as is often the case, your purpose is to test a hypothesis about a specific problem, you should state both the nature of the problem and what you expected to find. Your introduction should include the theory underlying the experiment and any pertinent background data or equations. You should refer to papers relevant to the experiment. It may be tempting to cite all of the papers that you've looked up and read, but it is best to select a few relevant examples to make your case. Save some of the other references for the *Discussion* section.

Many students find the introduction to be the easiest section to write. While the structure of a lab report is generally constrained, the structure of the introduction is fairly flexible. Here, you can use imagination and creativity to introduce sources and opinions. You can introduce controversies. You also have more freedom in terms of tense, as you can vary the tense of the introduction based on the content. For example, you could discuss a controversy or a quotation in the past tense, then switch to the future tense to present some aspect of the experiment that will be important in the future, and then switch to the present tense to build up an argument based on current sources of information.

If you decide to use a quotation, which is uncommon but perfectly allowable, you can reference almost any scientist or public figure, provided that the quote is appropriate to the experiment. A scientist, such as Darwin, is good for something venerable, but even a children's author, such as Lewis Carroll, is game: after all, where would we be in evolutionary theory without the Red Queen's statement, "It takes all the running you can do, to keep in the same place"? (Evolutionary biologist Leigh Van Valen used these words to formulate his Red Queen Hypothesis, a statement on the eternal arms race between organisms.)

In most scientific papers, the purpose appears at the end of the introduction, ideally as either a one-sentence hypothesis or statement of purpose. A common approach to writing an introduction is to think of it as a one-page essay that builds up to the statement of purpose. This mini-essay must include the ideas, concepts, and context necessary for the reader to understand the purpose of the experiment. Putting the purpose of the experiment in the final sentence also provides an excellent bridge to the following sections. Above all, your statement of hypothesis, objective, or purpose must be clear. Spend extra time clarifying this statement, because without clarity of purpose, the whole reason for doing the experiment could seem shaky.

To avoid common pitfalls,

- make sure your *Introduction* is not more than two pages in length;
- make your hypothesis and purpose as unambiguous as possible; and
- avoid anticipating your results.

Formulating a hypothesis

Before you write your introduction, you should understand the scientific method. In the life sciences, the scientific method has four basic steps. The first step is to consider experimental experience—both your own observations and the published observations of others. The second step is to formulate a conjecture or a hypothesis based on this experience. The third step is to combine the hypothesis with a prediction. The fourth step is to test the truth of this prediction. You must follow these steps when you are designing an experiment, as scientific advancement depends on a logical, repeatable sequence of actions. Once you have completed your experiment, others must be able to repeat your actions and, in turn, create their own hypotheses and predictions based on the most recent experience. These new hypotheses and predictions can then be tested by other researchers, and so on.

How do you formulate a hypothesis? One way is to follow the eight steps outlined below:

1. Start with a simple statement (e.g., "lead can cause poisoning").
2. Make the statement conditional by phrasing it as an *if/then* statement (e.g., "*if* lead causes poisoning, *then* people exposed to lead paint in their homes could suffer from lead poisoning").
3. Determine the variables (e.g., house lead content and number of cases of poisoned inhabitants). Note: an experiment to test a hypothesis requires at least two variables.
4. Determine the subject group (e.g., people who live in houses that have lead-painted surfaces).
5. Identify a treatment that will make the statement testable (e.g., exposure to lead paint in houses).
6. Consider the outcome (e.g., frequency of poisoning among occupants in houses with lead-painted surfaces).
7. Consider whether a control group (a non-treated subject group) is possible (e.g., people who live in houses that have never had any lead paint).

8. Put the elements together to form your hypothesis (e.g., "if lead causes poisoning, then people exposed to lead paint in their homes face a greater chance of poisoning than people who live in houses free of lead paint").

If you consider the example given in this list, you can see how it might be possible to have an introduction that has four paragraphs—the first discussing lead in paints, the second discussing widespread use of lead in house paint, and the third discussing toxicology of lead—followed by the final paragraph containing the hypothesis.

Materials and methods

The *Materials and Methods* (*M&M*) section contains a description of the materials and the equipment you used, some explanation of how you physically set up the experiment, an overview of the experimental design (treatments, controls, statistical tests and analyses, etc.), and a step-by-step description of your actions. This final element is essential, as readers must be able to repeat your experiment.

The *Materials* portion often lists the biological materials used in the experiment, such as the organisms sampled, purchased, or collected. Include the source and the full Latin name of any organism(s): for example, "California sea cucumbers, *Parastichopus californicus*, were purchased from a biological supply house (WestWind SeaLab Supplies, Victoria, BC)."

In most life sciences lab reports, it is very unusual to list or illustrate the experimental apparatuses with the biological materials. It is more common to include details of the lab equipment and chemicals in the *Methods* description. When larger pieces of equipment are mentioned, it is good practice to include the name and model, as well as the location of the manufacturer: for example, "LI-COR Leaf Area Meter LI-3100C (Lincoln, NE)." For chemicals, the name and location of the source or supplier are sometimes included: for example, "Acetonitrile (analytical grade purity) was obtained from Sigma-Aldrich, Oakville, ON."

In the following example from an experiment involving protein preparation, the materials and equipment are mentioned in the description of the methods:

> Protein samples were reduced with dithiothreitol (30 min at 37°C), and cysteine sulfhydryls were alkylated with iodoacetamide (30 min at 37°C in

darkness). Porcine trypsin (2 µg - Promega Corp., Madison, WI) was added to each sample, which was digested at 37°C for 16 h. The samples were desalted on a Waters Oasis HLB column (Waters Ltd., Mississauga, ON), vacuum-concentrated, and then stored at -80°C prior to LC-MS analysis.

In this example, there is no separate list or separate section for the materials. Specialized materials (e.g., porcine trypsin from Promega Corp.) and equipment (e.g., the chromatography column from Waters Ltd.) are referenced in more detail than common off-the-shelf materials (e.g., dithiothreitol and iodoacetamide). Also note that there is *no mention* of common lab equipment. In the first sentence, the protein samples that are being reduced are in tubes and vials that are not even mentioned. The temperature incubators in which the temperatures are maintained at 37°C are not mentioned. Why are these materials not included? Items that are considered standard are usually not mentioned. These standard items include lab equipment such as incubators, water baths, pH meters, refrigerators, glassware, and centrifuges, to name only a few.

For some lab reports you will need to include a simple diagram—produced on a computer or by hand—to help your reader visualize the arrangement of the equipment. For example, illustrations should be provided for specialized, custom-made equipment, such as a specialized trap to catch wolverines, or a custom-made glass manifold used for distributing gases to attached culture flasks. If the diagram is too large to fit on a regular page, you can label it and attach it at the end of the report. Just remember to refer to the attachment within the body of the report.

The *Methods* portion is a step-by-step description of how you carried out the experiment. Remember to describe the procedures in order. If your experiment consisted of a number of tests, you should begin your discussion of methods with a short summary statement listing the tests so that the reader will be prepared for the series. When you describe the tests in full, discuss them in the same order as they appear in this summary statement to avoid confusion.

If your experiment involves a series of processes, then it should be divided accordingly. For example, the first step in the experiment may consist of measurements repeated on different treatments. This may be followed by a data analysis step that involves statistical methods. The statistics should be described in a separate section, which follows the measurements and treatments section. If the experiments result in many illustrations—for example, if using scanning electron microscopy or fluorescence microscopy—then a separate section, such as one on microscopy, is warranted.

As you describe your methods, use enough detail so that others would have no difficulty repeating the experiment in all its essential details. If you are following instructions in a lab manual, you should not copy them out word for word—this might be considered plagiarism. In some cases, your lab instructor may permit you to simply refer to the instructions and add details of any deviation from the manual's instructions. If a certain procedure is long, complicated, or not necessary to a full understanding of the experiment, you may describe it in a labelled attachment at the end of the report.

Although you should be concise in your description of the experimental method, make sure that you don't omit essential details. From a scientist's standpoint, this section, more than any other in a lab report or scientific article, provides direct and immediate insight into the quality of the science being reported. A well-written, detailed, and clear *M&M* section provides a reader with confidence that the rest of the work is scientifically solid. In contrast, an *M&M* section that is missing details and appears to be unrepeatable raises alarm bells. Your lab report should include an appropriate level of detail that allows a person with a similar level of scientific expertise to repeat the experiment. Writing your report at a level that matches your scientific expertise explains why lab reports are likely to be more demanding as you progress through your years of study.

It takes some discernment to separate the useless details from the important ones. If you heated a test tube in a water bath, for example, you must report the temperature and the duration. In this case, the type of tube or water bath is immaterial to the outcome, whereas the temperature and duration are critical. If you performed a test or process at a faster or slower rate than usual, you must indicate the rate. Readers must know exactly what controls to apply if they try to perform the experiment themselves.

Tense and voice are generally invariable. The past tense is standard in all *M&M* sections. A lab report's methods section is not a cooking recipe. Cookbooks are normally written in the present, as if the writer were looking over the shoulder of an actively engaged reader. Scientific methods are written in the past tense, because this is a precise description of events that have already occurred.

In terms of voice, there has been some debate among scientists who write for scientific journals about whether to use the active or passive voice (e.g., "*I measured* the length of leaf petioles" versus "Leaf petioles *were measured*").

Thirty years ago, only the passive voice was used for this kind of writing because it emphasized the procedure rather than the person. However, the last two decades have seen a sharp rise in the tendency to use the active voice because it is clearer and less likely to produce awkward, convoluted sentences. Some researchers have also argued that third-person reporting is merely a pretense of objectivity—that is, using the first person is more honest, as it addresses the more modern sentiment that the scientist is not separable from his or her science. However, a section such as *Materials and Methods* is not about the experimenter so much as the experimental details. In many cases, *M&M* sections are best written in an impersonal voice. Ask your instructor or teaching assistant about his or her preferences, but also use your own judgment about what sounds best. Your goal, whichever voice you use, is to achieve clarity and objectivity.

To avoid common pitfalls,

- use lots of detail—it is all too easy to underwrite *M&M* sections—and
- remember to describe the controls.

Results

This is the section of most interest to a reader. It usually contains a mix of data, graphics, and verbal description. It will also likely contain some statistical analysis.

Instructors will explain whether they expect you to give the details of your calculations or only the results of those calculations. In either case, you should pay special attention to the units of any quantities; to omit or misuse them is a serious scientific mistake. Taking care to include all units will also reveal mistakes in your calculations that you might not have detected otherwise. Also, remember to use scientific notation when your calculations deal with numbers that are orders of magnitude higher or lower than the standard SI unit.

You should also make sure, where possible, that the calculated values you report include the "uncertainty" in each of them. For example, you might report that the calculated average diameter of the alga *Volvox* is 80.05 ± 0.35 μm. When reporting any calculations or measurements, check to see if you need to include either the standard deviation, the standard error of the mean, or the coefficient of variation. You should also keep in mind the difference between accuracy and precision: accuracy is a description of how close the values are to the true value(s), whereas precision is an indication of how repeatable the values are. If you get the same value over and over, it may be precise, but if it is far off the true or expected value, it may not be accurate.

The order of the *Results* section depends on the type of experiment you performed. If it is a relatively simple experiment, the section will begin with the main finding and then proceed to secondary findings—you will spark your reader's interest if you put the most important results first. If it is a complex experiment it is best described in the order in which the treatments were conducted.

Whenever possible, summarize your results in a graph or a table. A graph is usually preferable to a table because it has greater visual impact. However, if you have made several measurements, you might not be able to include your results in a single figure, in which case you are better off reporting them in tabular form. Software with sophisticated charting features are widely available that will allow you to create eye-catching figures quickly and easily (see Chapter 10 for a discussion of graphs and tables).

Graphs and tables summarize data, which means that this data should not be repeated in the body of the report. If the graph shows, for example, that after five years of growth the average height of fertilized poplar trees was 3.1 ± 0.3 m, as opposed to unfertilized trees that was only 2.2 ± 0.3 m, then you should not repeat these details in the text. When referring to this figure in the body of the report, it would be better to describe the trend—for example, "During the first five years of growth, fertilized poplar trees attained significantly greater heights than unfertilized trees (Fig. 1)." You have to assume that the reader gets to the point in the text when you refer to Figure 1 and that he or she then looks at your figure carefully before returning to the main text to continue reading. Repeating the data in the body of the text is clearly redundant, and redundancy is to be avoided in reporting as it gives the impression of unfinished, unpolished writing.

Whatever type of figure or table you use, label it clearly and be sure to refer to it and explain it in the text. Give each illustration a caption that includes a number and a title, and refer to it by number in your report. Figure captions appear below the figure, whereas table numbers and titles appear above the table. You should always number and present the figures in the order in which they are mentioned in the text. A common error is to position figures where they look best, rather than based on their chronological order—for example, Figure 5 must appear before Figure 6 even if it means reformatting some of your material. The caption should allow the figure to stand on its own. Do not simply write "Figure 1"; instead, describe it in full: "Figure 1: Graph of average lengths (cm ± SE) of maturing turtles versus time (wk). n = 5. Asterisks indicate significant differences determined by one-way ANOVA, at p < 0.01." If the illustration is a photo of two turtles and a dime (for scale), write, "Figure 1: Photograph of immature (left) and mature (right) turtles with a dime for comparison." Why are such full descriptions necessary? The reason is that figures and tables often do not fall on the same page on which they are referred to in the text. Figures and tables may appear together at the end of the report, or they may be incorporated in the body of the report. In the latter approach, the extent of the figures or tables may cause them to be separated by several pages from the point at which they are mentioned in the text. In such cases, having captions that are complete, stand-alone descriptions helps readers to keep track of information as they flip back and forth through your report.

Graphs can be highly effective for displaying information, but they also require a lot of planning. Remember the following guidelines when you are creating graphs for your lab report:

- Use a scale that will allow you to distribute your data points as widely as possible on the page.
- Put the independent variable (the one you manipulated) on the horizontal axis and the dependent variable (the one you measured) on the vertical axis.
- Make the vertical axis approximately three-quarters the length of the horizontal axis.
- Use large and distinctive symbols, with different symbols for each line on the graph.
- Put error bars (±) on data points where known.
- Label the axes clearly and always include the units of measurement so that the reader knows exactly what you have plotted on the graph.
- Include a legend, when necessary, to indicate the different units and to explain what the different symbols represent.
- Remember to give the graph a title and a caption—position the caption below the graph.

In most cases, you will use the past tense to discuss your results, as you are reporting the outcome of a completed experiment. As in the *M&M* section, you may choose to use the active or passive voice where appropriate. Do not include any references in the *Results* section. You should use references in the

Introduction, *Materials and Methods*, and *Discussion*, but never in the *Abstract* or *Results* sections.

To avoid common pitfalls,

- do not repeat data in the text when it is illustrated in the tables and figures,
- write this section so that it can be read aloud—it should not sound like a lab book, and
- do not use point form.

Discussion

The *Discussion* section of the lab report allows you the greatest intellectual freedom because it is here that you analyze and interpret the test results and comment on their significance. In this section, you should state whether the test produced its predicted outcome or if the results were unexpected. It is important to discuss those elements that influenced the results. For a good discussion, remember to think critically not only about your own work but also about how it relates to previously published work. In determining what details to include in your analysis, you might try to answer the following questions:

- Do the results reflect the hypothesis or purpose of the experiment?
- Do the results agree with previous findings as reported in the literature on the subject? If not, can you account for the discrepancy between your own data and the values accepted or obtained by other students and scientists?
- Could the results have another explanation?
- Did the procedures you used help you accomplish the purpose of the experiment? Does your experience in this experiment suggest a better approach for next time?
- What (if anything) may have gone wrong during your experiment and why? What are the sources of error and are they included in the statistical method of analysis?

If you do find errors in your experiment, try to break them down into instrument errors (e.g., due to an improperly calibrated measuring device), method errors (e.g., due to uncontrolled elements, such as room temperature, that varied from batch to batch), and finally—every student's familiar foe—human error. If this is your error, admitting as much is good, but describing how the error came about is even better, as it shows that you understand how a particular process should occur. This is not just a question of honesty, it is a necessary step in acquiring the introspection required to be a good scientist. Undergraduates are usually paired with a lab partner. If the error is due to the lab partner, it is prudent to be diplomatic and avoid accusing your lab partner of incompetence. Mistakes happen. They should be noted, but it is inappropriate to use a scientific report to pass judgment on personal matters. There is a silver lining to constant error analysis. Every scientist can relate a personal story in which analysis of errors led directly to a discovery.

The structure of the discussion should follow a logical progression. In the first paragraph of the *Discussion* section, you should always discuss the most important result and whether the hypothesis is true or false. This should be followed by paragraphs that position the results in terms of scientific literature. For example, if the experiment is about some aspect of animal physiology, you would discuss whether your hypothesis was proved or disproved in the first paragraph. Then you would highlight your most important result. In the next paragraph, your results would be discussed in the context of relevant physiological literature. If the type of animal studied was important, for example if the species exhibited a peculiar physiological response, then the evolutionary context would need to be discussed. The directions that the discussion can take partly depend on how well you understand the literature. More subtle interpretations become possible the more you have read and understood.

This is where reading pays off. If you assembled the information from your readings in an accessible manner, you will be able to structure your arguments more effectively. Information from scientific articles can be summarized in different ways. If only a few papers are needed for a short lab report, then well-annotated articles will suffice. For longer lab reports, annotated bibliographies provide excellent sources from which to extract relevant information. Organizing summarized material by subject will often provide an obvious order for subject paragraphs within the discussion.

You do not need to write the *Discussion* section in the past tense, unlike the *M&M* and *Results* sections. This is because you are considering not only past scientific literature and historically important experiments but also current research publications and modern trends in the field. You can avail yourself of the present tense when it seems appropriate. Furthermore, the future tense is used when recommending research areas for future experiments.

Avoid referring to figures or tables in this section. Why? One answer is that the *Discussion* section is about the *implications* of the data, not the data itself—the

Results section is the place to describe and note values. You may be tempted to say, "The data in Figure 1 compares favourably with that of Davidson and co-workers (2018)," but it is better to write, "Our sample of maturing snapping turtles show length increases similar to a recent study of the same species (Davidson et al. 2018)." Another reason not to discuss individual values is that your reader will be reading your report all at once and does not need reminding about the relevant important data.

No matter how frustrating the lab experience has been, do not begin with a discussion of the problems you faced—they have little to do with the purpose of the experiment. Poor results are usually the fault of either poor experimental design or poor execution, not of a poor hypothesis. You should examine difficulties in experimental execution in the body of the *Discussion* rather than at the beginning or end, where they leave a distinctly bitter impression. Similarly, you should bury your suggestions for improvements to the experiment in the middle.

The *Discussion* section should end on a strong note. Try and make a statement that emphasizes the importance of your results—for example, "In this experiment, we showed that the ideal pH range during isolation of white spruce protoplasts was between pH 5.5 and 6.5, which has practical applications for technologies that depend on large-scale yield of protoplasts, such as bioreactor technology." Generally, you can end your *Discussion* section in one of two ways. If your instructor has requested a formal *Conclusions* section, then you should end your *Discussion* with a brief summary. Otherwise, you can end the *Discussion* section with a final paragraph that presents both conclusions and future perspectives.

To avoid common pitfalls,

- spend enough time editing this section to make your ideas as clear as possible,
- write dynamically and avoid using one long sentence after another, and
- use a punch finish—do not let your discussion drift to its death.

Conclusions

The *Conclusions* section is a brief statement of the conclusions that you have drawn from the experiment. You don't necessarily need a separate section for your conclusions; they can also appear as a short summary paragraph at the very end of the *Discussion* section. You may include a chart (or table or graph) if you think it will better model and clarify the conclusion.

References

The only way to avoid suspicion of plagiarism is to support every non-original statement with a reference citation. Each time you refer to a book or an article in the text of your report, cite the reference; then, at the end of the paper, make a list of all the sources you have cited. The precise format of the citations and reference list varies among disciplines, so you should check with your instructor or TA to see which style you should use. For details on the correct form for scientific documentation, see Chapter 11.

Some sections in a lab report require references, while others are always free of references. You will include lots of references that support your line of reasoning in the *Introduction* and *Discussion* sections. In the *M&M* section, you may need to reference methods that you found in the literature if they are necessary for the reader to understand how to repeat the experiment. Though common, such references are usually few in number. However, *never* include even a single reference in the *Abstract* or *Results* sections. Some students are surprised by this last convention, but, after all, your results are your own and you must describe them, not refer to someone else's results. Including outside work constitutes a comparison, which would, logically enough, turn a *Results* section into a *Discussion* section.

To avoid common pitfalls,

- follow the citation style for your discipline;
- make sure you haven't used references in the wrong sections; and
- check your references line-by-line to make sure you haven't missed anything.

Attachments or appendices

In some cases you may wish to include various raw data, supplementary photos, and detailed calculations as attachments or appendices. These additions should be placed on separate pages at the end of the report.

Putting the Sections Together

Should you write the sections of your report in order? No!

It is far, far easier to write these sections out of order. No professional scientist writes straight through from the *Abstract* to *Conclusions*. What follows is a recommended order for writing your lab reports:

- Begin by writing the *Materials and Methods* section, which is often the easiest to compose because of the strictly descriptive nature.
- Then, turn to the *Results* section, which also tends to be straightforward because it involves graphic interpretation and description of analyses.
- Next, write your *Introduction*, which requires bolstering the purpose and rationale of your experiment with examples from the scientific literature. Writing this section at this point in the process will also remind you of the important aspects of the experiment's goals.
- Write the *Discussion*, where you can elaborate on the importance of your results within the context of the primary literature, some of which you will have referenced in the *Introduction*.
- After the *Discussion*, write your brief *Conclusions*.
- Once you have prepared these sections, you will be ready to write a succinct summary for the most difficult of all the sections—the *Abstract*.

Summary

Lab reports are the most common writing assignments you will face as an undergraduate in the life sciences. These reports follow formal conventions found in scientific papers, so you must learn to write within these conventions in order to publish as a scientist. Lab reports are traditionally divided into seven sections: *Abstract, Introduction, Materials and Methods, Results, Discussion, Conclusions,* and *References*. Each of these sections has specific content requirements and stylistic constraints.

The *Abstract* is the most specific, and you can learn the appropriate style by reading and imitating published journal abstracts. The *Introduction* and *Discussion* sections must tie together—one ends where the other begins. The *Introduction* is where you can make arguments that build to support the logic of the hypothesis or purpose of the experiment. The *Discussion* is where you can explore whether you have proved or achieved this same hypothesis or purpose. The *Materials and Methods* section is fairly straightforward, but you have to be careful to include every detail relevant to the experiment. The *Results* section requires a clear, illustrated version of the data you recorded during the experiment and a clear, written description of this data. Your conclusions, whether they appear in their own section or at the end of the *Discussion* section, should be brief and highlight the most important implications of your experiment. Finally, the references must be listed correctly—edit them slowly and carefully to achieve a professional finish.

6 Writing with Style

Objectives
- Finding your voice
- Achieving clarity
- Choosing your words carefully
- Controlling clauses, sentences, and paragraphs
- Writing with force and purpose

Writing with style does not mean inflating your prose with fancy words and extravagant images. Any style, from the simplest to the most elaborate, can be effective depending on the occasion and intent. Writers known for their style are those who project their own personality into their writing; we can hear a distinctive voice in what they say. Obviously it takes time to develop a unique style. To begin with, you have to decide what general effect you want to create.

Taste in style reflects the times. In earlier centuries, many respected writers wrote in an elaborate style that we would consider much too wordy. Today, journalists have led the trend towards short, easy-to-grasp sentences and paragraphs. Writing in an academic context, you may expect your audience to be more reflective than the average newspaper reader, but the most effective style is still one that is clear, concise, and forceful.

Learning to find your voice in writing is important, and achieving clarity does not mean eliminating your personality. Occasionally, a student expresses the mistaken belief that there must be a universal, neutral style common to the sciences. The more you read scientific literature, the more styles and voices you will encounter. Scientific peer-reviewed journal articles appear to have a flat, just-the-facts style; in part, this is because scientific journals have a captive readership—if you are in the field, you should be reading the article. This is the style that your professors and TAs would like you to emulate. In comparison to journal articles, review articles can be highly individualistic. Even more individual expression is evident in articles in lay science magazines—for example, *New Scientist* features articles that are written to capture readers' attention. These types of publications depend on sales to the public and have to be written in an engaging style. A concession made by scientific journals in the last three decades is one that is reflected in sciences in general: greater acceptance of first-person narration. Getting a handle on when to use *I* and *we* takes time, but it is worth mastering.

Be Clear

Use clear diction

Vocabulary choice should be precise. A good dictionary, online or in print, will help you understand unfamiliar words and archaic and technical senses of common words. Some dictionaries will help you use words properly by offering example sentences that show how certain words are typically used. A dictionary will also help you with questions of spelling and usage. If you aren't sure whether a particular word is too informal for your writing or if you have concerns that a certain word might be offensive, consult a good dictionary.

You should be aware that Canadian usage and spelling may follow either British or American practice but usually combines aspects of both. There are a number of Canadian dictionaries available online, as well as in print, that will help you to be consistent in your approach. It's also a good idea to make sure that the *Language* feature of your word-processing program is set to *English (Canada)*.

A thesaurus lists words that are closely related in meaning. It can help when you want to avoid repeating yourself or when you are fumbling for a word that's on the tip of your tongue. Your word-processing program also has a thesaurus feature that allows you to look up synonyms and antonyms easily. Be careful, though: make sure you distinguish between denotative and connotative meanings. A word's *denotations* are its primary (or "dictionary") meanings. Its *connotations* are any associations that it may suggest; they may not be as exact as the denotations, but they are part of the impression the word conveys. If you examine a list of synonyms in a thesaurus, you will see that even words with similar meanings can have dramatically different connotations. For example, alongside the word *watery*, your thesaurus may give the following entries: *anemic, aqueous, colourless, diluted, insipid, marshy, muddy, soggy, waterlogged*, and *weak*, to list but a few of the possibilities. Imagine the different impressions you would create if you chose one or the other of those words to complete this sentence: "The extraction of the zebra lymph tissue was done according to the protocol, but the resulting fluid was far more _____ than previous extractions from horse lymph tissue." In order to write clearly, you must remember that a reader may react to the suggestive meaning of a word as much as to its "dictionary" meaning.

Use plain English

A side effect of having to master heaps of life sciences terminology and field-specific jargon is that it inspires insipid prose. You should use jargon only when it is required for plain illustration. A high concentration of technical jargon may give you an increased sense of sophistication, but you risk confusing your readers by forcing them to decipher and remember complex terminology when they should be focusing on your arguments. Another unfortunate effect of using jargon is that it tempts young life sciences writers into using fancy words and phrases. It can also tempt writers to use unnecessary or unclear abbreviations:

orig. EtOH was used to sterilize the hood.
rev. Ethanol was used to sterilize the laminar flow hood.

Plain words are almost always more forceful than fancy ones. If you aren't sure what plain English is, think of the way you talk to your friends (apart from swearing and slang). Many of our most common words—the ones that sound most natural and direct—are short. A good number of them are Anglo-Saxon in origin and are among the oldest words in the English language. In contrast, most of the words that English has derived from other languages are longer and more complicated; even those that have been used for centuries can sound artificial. For this reason you should beware of words loaded with prefixes (*pre-*, *post-*, *anti-*, *pro-*, *sub-*, *maxi-*, etc.) and suffixes (*-ate*, *-ize*, *-tion*, etc.). These Latinate attachments can make individual words more precise and efficient, but putting a lot of them together will make your writing seem dense and hard to understand. In many cases you can substitute a plain word for a fancy one.

Fancy	**Plain**
accomplish	do
cognizant	aware
commence	begin, start
conclusion	end
determinant	cause

fabricate	build
finalize	finish, complete
firstly	first
maximization	increase
modification	change
numerous	many
obviate	prevent
prioritize	rank
requisite	needed
sanitize	clean
subsequently	later
systematize	order
terminate	end
transpire	happen
utilize	use

Suggesting that you write in plain English does not mean that you should never pick an unfamiliar word or a foreign derivative; sometimes those words are the only ones that will convey precisely what you mean. Inserting an unusual expression into a passage of plain writing can also be an effective means of catching the reader's attention—as long as you don't do it too often.

Be precise

Always be as specific as you can. Avoid all-purpose adjectives such as *major*, *significant*, and *important* and vague verbs such as *involve*, *entail*, and *exist* when you can be more specific:

> **orig.** The method was significantly bettered by pre-treatment of slides with poly-L-lysine.
>
> **rev.** The method was improved by pre-treatment of slides with poly-L-lysine.

Here's another example:

> **orig.** Exploration of the fossil cliffs at Joggins, on the Bay of Fundy, led to major discoveries in paleoecology.
>
> **rev.** Exploration of the fossil cliffs at Joggins, on the Bay of Fundy, led to revolutionary discoveries in paleoecology.

Avoid unnecessary qualifiers

Qualifiers such as *very, rather,* and *extremely* are overused. Saying that something is *very complex* may have less impact than saying simply that it is *complex*. An easy way to eliminate this problem in your writing is to type a word (e.g., *very*) into your word processor's *Find* function and then eliminate or replace each one. For example, compare these sentences:

> The voodoo lily is notorious for its very repulsive smell.
> The voodoo lily is notorious for its strikingly repulsive smell.

Which has more impact? When you think that an adjective needs qualifying—and sometimes it will—first see if it's possible to change either the adjective or the phrasing. Instead of writing

> The microtome required a very big change in alignment to accommodate large specimens,

write a precise statement:

> The microtome required substantial realignment to accommodate large specimens,

or (if you aren't sure whether or not the alignment is the key element):

> The microtome had to be set up to accommodate large specimens.

In some cases, qualifiers not only weaken your writing but are redundant because the adjectives themselves are absolutes. To say that something is very unique makes as little sense as saying that someone is slightly pregnant or extremely dead.

Create clear paragraphs

Paragraphs come in so many sizes and patterns that no single formula could possibly cover them all. The two basic principles to remember are these:

1. a paragraph is a means of developing and framing an idea or impression; and
2. divisions between paragraphs aren't random but indicate a shift in focus.

Develop your ideas
You are not likely to sit down and consciously ask yourself, "What pattern shall I use to develop this paragraph?" What comes first is the idea you intend to develop; the structure of the paragraph should flow from the idea itself and the way you want to discuss or expand it.

You may take one or several paragraphs to develop an idea fully. For a definition alone you could write one paragraph or ten, depending on the complexity of the subject and the nature of the assignment. Just remember that ideas need development, and that each new paragraph signals a change in idea.

Consider the topic sentence
Skilled skim readers know that they can get the general drift of a book or an article simply by reading the first sentence of each paragraph. The reason is that most paragraphs begin by stating the central idea to be developed. In a formal essay, each topic sentence should relate to the section in which the paragraph appears.

Like the thesis statement for the essay as a whole, the topic sentence is not obligatory; in some paragraphs you might not state the controlling idea until the middle or even the end, and in others you might not state it at all but merely imply it. Nevertheless, it's a good idea to think out a topic sentence for every paragraph. That way you'll be sure that each one has a readily graspable point and is clearly connected to what comes before and after. When revising your initial draft, check to see that each paragraph is held together by a topic sentence, either stated or implied. If you find that you can't formulate one, you should probably rework the whole paragraph.

Maintain focus
A clear paragraph should contain only those details that are in some way related to the central idea. It should also be structured so that the details link naturally to one another. One way of showing these relationships is to keep the same grammatical subject in most of the sentences that make up the paragraph. When the grammatical subject keeps shifting, a paragraph loses focus, as in the following example:

> ***orig***.   Cell membranes have been the subject of much research for many decades. In 1972, Singer and Nicolson found that these membranes are composed of a lipid bilayer in which proteins are

randomly distributed. Within a few years, <u>scientists</u> discovered that biomembranes have heterogeneous regions called microdomains or rafts. <u>Clusters</u> of sphingolipids and cholesterol typify these zones. <u>Biochemists</u> have been able to create artificial membranes that replicate these regions. <u>Microdomains</u> remained a peripheral idea, until glycosphingolipid clustering in Golgi membranes was found to occur just before vesicles are released. This new <u>information</u> was used to frame the lipid raft hypothesis of Simons and Ikonen in 1997. The <u>localization</u> of particular signaling proteins in portions of membranes enriched in particular lipids was an important evolutionary step in the development of signal transduction for both prokaryotes and eukaryotes (Bramkamp and Lopez, 2015).

Here the grammatical subject (underlined) changes from sentence to sentence. Notice how much stronger the focus becomes when all the sentences have the same grammatical subject—either a variation on the same noun, a synonym, or a related pronoun:

rev. <u>Researchers</u> have studied cell membranes for many decades. Early on, <u>scientists</u> discovered that these membranes are composed of a lipid bilayer in which proteins are randomly distributed (Singer and Nicolson 1972). Later, <u>biochemists</u> discovered that these biomembranes have heterogeneous regions, called microdomains or rafts, characterized by high concentrations of sphingolipids and cholesterol. <u>Biochemists</u> have been able to create artificial membranes that replicate these regions. Subsequently, <u>they</u> have discovered Golgi membranes with regions of high glycosphingolipid concentration just prior to vesicle formation. <u>Simons and Ikonen</u> (1997) used this information to form the lipid raft hypothesis of membrane structure. Today, <u>cell biologists</u> agree that localization of particular signaling proteins in portions of membranes enriched in particular lipids was an important evolutionary step in the development of sophisticated signal transduction in both prokaryotes and eukaryotes (Bramkamp and Lopez, 2015).

Naturally it's not always possible to retain the same grammatical subject throughout a paragraph. If you were comparing topics from different fields of study, for example, you might have to switch back and forth. In the same way, you have to shift when you are discussing examples of an idea or exceptions to it.

Avoid monotony

If most or all of the sentences in your paragraph have the same grammatical subject, how do you avoid boring your reader? There are two easy ways:

1. **Use substitute words.** Pronouns—including personal (*I*, *we*, *you*, *he*, *she*, *it*, *they*), demonstrative (*this*, *that*, *these*, *those*), and indefinite (*someone*, *everyone*, *many*, *others*, etc.) pronouns—can replace the subject, as can synonyms (words or phrases that mean the same thing). The revised paragraph on cell membranes, for example, includes the pronouns *they* and *others*, which substitute for *researchers*. Most well-written paragraphs have a liberal sprinkling of these substitute words.
2. **Highlight the subject by juxtaposing it against an opening clause, phrase or word.** When the subject is placed in the middle of the sentence rather than at the beginning, it's less obvious to the reader. If you take another look at the revised paragraph, you'll see that in a few sentences there is a word or phrase in front of the subject—*early on*, *later*, *subsequently*, *more recently*. As you can see, even a single word will do the trick.

Link your ideas

To create coherent paragraphs, you need to link your ideas clearly. Linking words are those connectors—conjunctions and conjunctive adverbs—that show the relationship between one sentence, or part of a sentence, and another. They're also known as transition words, because they form a bridge from one thought to another. Make a habit of using linking words when you shift from one grammatical subject or idea to the next, whether the shift occurs within a single paragraph or as you move from one paragraph to another. Here are some of the most common connectors and the logical relations they indicate:

Linking word	Logical relation
and	
also	
again	
furthermore	addition to previous idea
in addition	
likewise	
moreover	
similarly	
alternatively	
although	
but	
despite, in spite of	
even so	
however	change from previous idea
in contrast	
nevertheless	
on the other hand	
rather	
yet	
accordingly	
as a result	
consequently	
hence	summary or conclusion
for this reason	
so	
therefore	
thus	

Numerical terms such as *first*, *second*, and *third* also work well as links.

A few words on the above list pose problems for students. One example is *however*. This word is often used incorrectly as a conjunction (it is actually a conjunctive adverb; see the discussion of commas in Chapter 8 for more on conjunctions and conjunctive adverbs). To add to the woes, *however* can mean different things depending on its place in a sentence. When used at the beginning of a sentence but not followed by a comma, it means "in whatever way":

> However you conduct this experiment, the electrodes will malfunction.

But *however* can also mean "nevertheless." When used in this second sense, its placement can vary depending on what the writer wants to emphasize, but it must be restricted by commas:

> However, cold temperatures affected electrode function.
> Cold temperatures, however, affected electrode function.
> Cold temperatures affected, however, electrode function.

Vary paragraph length, but avoid extremes
Ideally, academic writing will have a balance of long and short paragraphs. However, it's best to avoid the extremes—especially the one-sentence paragraph, which can only state an idea without explaining or developing it. A series of very short paragraphs is usually a sign that you have not developed your ideas in enough detail or that you have started new paragraphs unnecessarily. On the other hand, a succession of long paragraphs can be difficult to read. In deciding when to start a new paragraph, consider what is clearest and most helpful for the reader.

Be Concise

Strong writing is always concise. It leaves out anything that does not serve some communicative or stylistic purpose, and it says as much as possible in as few words as possible. Concise writing will help you do better on both your essays and your exams.

Use adverbs and adjectives sparingly

Don't sprinkle adverbs and adjectives everywhere and don't use combinations of modifiers unless you are sure they clarify your meaning. One well-chosen word is always better than a series of synonyms:

> **orig.** As well as being costly and financially extravagant, the field trial is reckless and risky.
> **rev.** The field trial is risky as well as costly.

Another problem with adverbs in life sciences essays and reports is that they generally serve as woolly filler. *Interestingly, importantly, approximately, definitely, absolutely, roughly,* and *fairly* are only a few examples of the kinds of adverbs that you should always eliminate in favour of more concise descriptions or arguments.

Avoid noun clusters

A recent trend in some writing is to use nouns as adjectives (as in the phrase *noun cluster*). This device can be effective occasionally, but frequent use can produce a mess. Noun clusters have a predictable "right-handedness," with the important noun always being the one furthest to the right. Piling up nouns in front of this one is like building with Lego. Unfortunately, the brevity achieved does not result in greater clarity. Breaking up noun clusters may not always result in fewer words, but it will make your writing easier to read:

> ***orig.*** experiment plan revision summary
> ***rev.*** summary of the revised experiment plan

Many noun clusters have become commonly accepted in the life sciences. While you won't be able to entirely avoid these clusters, you should avoid cluttering your writing with new noun combinations.

Avoid chains of relative clauses

Sentences full of clauses beginning with *which, that,* or *who* are usually wordier than necessary. Try reducing some of those clauses to phrases or single words:

> ***orig.*** The sugar analysis <u>that</u> was produced by the working group has a practical benefit, <u>which</u> was easily grasped by managers who have no technical training.
> ***rev.*** The sugar analysis produced by the working group has a practical benefit, easily grasped by non-technical managers.

Try reducing clauses to phrases or words

Independent clauses can often be reduced by subordination. Here are a few examples:

orig.   The report was written in a clear and concise manner, and it was widely read.
rev.   Written in a clear and concise manner, the report was widely read.
rev.   Clear and concise, the report was widely read.

orig.   The genotypes of wheat were planted in blocks, and the blocks were surrounded by buffer zones.
rev.   The wheat genotypes were planted in blocks surrounded by buffer zones.

Eliminate clichés and circumlocutions

Trite or roundabout phrases may flow from your pen automatically, but they make for stale prose. Unnecessary words are deadwood; be prepared to slash ruthlessly to keep your writing vital:

Wordy	*Revised*
due to the fact that	because
at this point in time	now
consensus of opinion	consensus
in the near future	soon
when all is said and done	[omit]
in the eventuality that	if
in all likelihood	likely
it could be said that	possibly, maybe
in all probability	probably

Avoid lazy beginnings

Although it may not always be possible, try to avoid beginning sentences with *It is . . .* or *There is (are). . . .* Your sentences will be crisper and more concise:

orig.   <u>It is</u> rare for marine biologists to see giant squid in Nova Scotia.
rev.   Marine biologists rarely see giant squid in Nova Scotia.

Lazy writing is pretty easy to detect. Too many sentences starting with either *this*, *that*, or *those* lend a paragraph a certain dullness. If you keep a weather eye on these words, you will gain much better control over your writing.

Be Forceful

Developing a forceful, vigorous style simply means learning some common tricks of the trade and practising them until they become habit.

Choose active over passive verbs

An active verb creates more energy than a passive one does:

> **Active:** The lion ate the gazelle.
> **Passive:** The gazelle was eaten by the lion.

Moreover, passive constructions tend to produce awkward, convoluted phrasing. Writers of bureaucratic documents are among the worst offenders:

> **orig.** It has been decided that the utilization of small rivers in the province for purposes of generating hydroelectric power should be studied by our department and that a report to the deputy minister should be made by our director as soon as possible.

The passive verbs in this mouthful make it hard to tell who is doing what. Passive verbs are appropriate in four cases:

1. When the subject is the passive recipient of some action:

    > The oxygen meter was calibrated by the senior lab technician.

2. When you want to emphasize the object rather than the person acting:

    > The antipollution policy in all three jurisdictions will be improved.

3. When you want to avoid an awkward shift from one subject to another in a sentence or paragraph:

    > We developed our understanding of the nitrogen cycle by studying terrestrial environments but were overwhelmed by new information when cold seeps in the deep oceans yielded their secrets.

4. When you want to avoid placing responsibility or blame:

   Several errors <u>were made</u> in the calculations.

When these exceptions don't apply, make an effort to use active verbs for a livelier style.

Use personal subjects

Most of us find it more interesting to learn about people than about things. Wherever possible, therefore, make the subjects of your sentences personal. This trick goes hand in hand with the use of active verbs. Almost any sentence becomes livelier with active verbs and a personal subject:

orig. The <u>consequence</u> of Haldane's experiment <u>was</u> the <u>decision</u> to select another species.
rev. After the experiment, <u>Haldane decided</u> to select another species.

Here's another example:

orig. <u>It is assumed</u> that few eagles died in the storm because a high number of eagles <u>were recorded</u> eight months later during the annual bird count.
rev. <u>We assume</u> that few eagles died in the storm because <u>we recorded</u> a high number of eagles eight months later during the annual bird count.

Use concrete details

Concrete details are easier to understand—and to remember—than abstract theories. Whenever you are discussing abstract concepts, therefore, always provide specific examples and illustrations; if you have a choice between a concrete word and an abstract one, choose the concrete. Consider this sentence:

orig. La Pérouse sailed across the Pacific and collected scientific information.

Now see how a few specific details can bring the facts to life:

> *rev.* La Pérouse sailed his two ships, the *Boussole* and the *Astrolabe*, across the Pacific and, like his admired predecessor Captain Cook, collected biological, ethnographic, and geographic information.

Adding concrete details doesn't mean getting rid of all abstractions. Just try to find the proper balance. The above example is one instance where you can improve your writing by adding words, as long as they are concrete and correct.

Make important ideas stand out

Experienced writers know how to manipulate sentences in order to emphasize certain points. The following are some of their techniques.

Place key words in strategic positions

The positions of emphasis in a sentence are the beginning and, above all, the end. If you want to bring your point home with force, don't put the key words in the middle of the sentence. Save them for the end:

> *orig.* Humans are less afraid of losing polar bears than of losing an entire ecosystem in this era of global climate change.
>
> *rev.* In this era of global climate change, humans are less afraid of losing polar bears than of losing an entire ecosystem.

Subordinate minor ideas

A common stylistic mistake in first drafts of reports and essays is connecting incidents with a string of *and*s, which has the effect of making all incidents of equal importance:

> *orig.* The pH was reached, <u>and</u> the titration was carried out, <u>and</u> the solution's colour changed.

Subordination of minor ideas in one part of a sentence allows the writer to emphasize another point:

> *rev.* Once the pH was reached, the titration was run until the solution's colour changed.

Major ideas stand out more and connections become clearer when minor ideas are subordinated:

orig. The class members gathered preliminary data, and they adjusted the experimental design.
rev.   After they gathered preliminary data, the class members adjusted the experimental design.

Make your most important idea the focus of the main clause, and try to put it at the end, where it will be most emphatic:

orig. The fish fry were dying due to sea lice.
rev.   Due to sea lice, the fish fry were dying.

Vary sentence structure

As with anything else, variety adds spice to writing. One way of adding variety that will also make an important idea stand out is to use a periodic rather than a simple sentence structure.

Most sentences follow the simple pattern of subject–verb–object (plus modifiers):

Many people eat meat.
     S   V   O

A *simple sentence* such as this gives the main idea all at once, in a single clause, and therefore creates little tension. A *periodic sentence*, on the other hand, does not give the main clause until the end, after one or more subordinate clauses:

Although studies have shown the environmental consequences of widespread animal farming, and although there are other efficient sources of protein, many people eat meat.
                                          S   V   O

The longer the periodic sentence is, the greater the suspense and the more emphatic the final part. Because this high-tension structure is more difficult to read than the simple sentence, your reader would be exhausted if you used it too often. Save it for those times when you want to make a very strong point.

Vary sentence length

A short sentence can add impact to an important point, especially when it comes after a series of longer sentences. This technique can be particularly useful for conclusions. Don't overdo it, though—a string of long sentences may be monotonous, but a string of short ones can make your writing sound like a children's book.

Still, academic papers usually have too many long sentences rather than too many short ones. Because short sentences are easier to read, try breaking up clusters of long ones. Check any sentence over 20 words or so in length to see if it will benefit from being split.

Use contrast

You can highlight an idea by placing it against a contrasting background:

> **orig.** Most sea mammals do not have toenails.
> **rev.** Unlike sea cows, most sea mammals do not have toenails.

Using parallel phrasing will increase the effect of the contrast:

> Although bromeliads tolerate drought, salt spray, and heat, they cannot tolerate cold.

Use a well-placed adverb or correlative construction

Adding an adverb or two can sometimes help you dramatize or clarify a concept:

> **orig.** Although this method is reliable, at the ambient desert temperatures it was fickle.
> **rev.** Although this method is <u>usually</u> reliable, at the ambient desert temperatures it was <u>always</u> fickle.

Correlatives such as *both . . . and* or *not only . . . but also* can be used to emphasize combinations as well:

> **orig.** The mass spectrometer separates ions by mass and by charge.
>
> **rev.** The mass spectrometer separates ions <u>not only</u> by mass <u>but also</u> by charge.

(or)

> **rev.** The mass spectrometer separates ions <u>both</u> by mass <u>and</u> by charge.

Use repetition
Repetition is a highly effective device for adding emphasis:

> According to E.O. Wilson, the inventor of sociobiology, humans need an evolutionary epic—an epic of creation, an epic of our beginnings, an epic of our place in the world—that would be as ennobling as any religious epic.

Of course, you would only use such a dramatic technique on rare occasions.

Some Final Advice

Write before you revise

No one expects you to sit down and put all this advice into practice as soon as you start to write. You would feel so constrained that it would be hard to get anything down on paper at all. You will be better off if you begin concentrating on these guidelines during the final stages of the writing process when you are looking critically at what you have already written. Some experienced writers can combine the creative and critical functions, but most of us find it easier to write a rough draft first before starting the detailed task of revising and editing.

Use your ears

Your ears are probably your best critics; make good use of them. Before producing a final copy of any piece of writing, read it out loud in a clear voice. The difference between cumbersome and fluent passages will be unmistakable.

Summary

The facts rarely speak for themselves—they usually need a writerly helping hand. As you face a highly stylized writing assignment, such as a lab report or an essay, you should learn to depend on the same skills that all writers use. Remember that your goal is to provide clarity. Choose appropriate words, compose forceful sentences, and create dynamic paragraphs to support your line of reasoning. Always pay attention to adverbs, adjectives, sentence lengths, parallel structures, comparisons, and rhetorical devices such as repetition—they can all help you keep your reader interested in your writing. The benefit of learning to control style is that you will begin to write with individuality—you will find your voice. This newfound ability to express yourself will make your experience of the life sciences more personal and, not surprisingly, more creative.

7 Common Errors in Grammar and Usage

Objectives
- Maintaining sentence unity
- Keeping subjects and verbs in agreement
- Using proper verb tenses
- Avoiding difficulties with pronouns
- Understanding modifiers
- Keeping order among pairs and parallels

This chapter is not a comprehensive grammar lesson; it's simply a survey of those areas where students most often make mistakes. It will help you pinpoint weaknesses as you edit your work. Once you get into the habit of checking your work, it won't be long before you are correcting potential problems as you write.

The grammatical terms used here are the most basic and familiar ones; if you need to review some of them, see the glossary. If you're interested in a more exhaustive treatment, consult one of the many books that deal exclusively with grammar and usage.

Sentence Unity

Sentence fragments

To be complete, a sentence must have both a subject and a verb in an independent clause; if it doesn't, it's a fragment. There are times when it is acceptable to use a sentence fragment in order to give emphasis to a point:

✔ Will the government pass legislation to prevent trophy hunting? <u>Not likely</u>.

Here the sentence fragment *Not likely* is clearly intended to be understood as a short form of *It is not likely that the government will pass this legislation.* Unintentional sentence fragments, on the other hand, usually seem incomplete rather than shortened:

> ✗ The Vancouver Island marmot is found exclusively in Canada. <u>Being an endemic species</u>.

The last "sentence" is incomplete because it lacks an independent clause with a subject and a verb. Remember that a participle such as *being* is a verbal, or "part-verb," not a verb. You can make a complete sentence by joining the fragment to the preceding sentence:

> ✔ Being an endemic species, the Vancouver Island marmot is found exclusively in Canada.

Run-on sentences

A run-on sentence is one that continues beyond the point where it should have stopped:

> ✗ Mosquitoes and blackflies are annoying, but they don't stop biologists from gathering data in the field, and such is the challenge in the Arctic.

This run-on sentence could be fixed by adding a period or a semicolon after *in the field* and removing the word *and*.

Another kind of run-on sentence is one in which two independent clauses are wrongly joined by a comma. An independent clause is a phrase that can stand by itself as a complete sentence. Two independent clauses should not be joined by a comma without a coordinating conjunction:

> ✗ The gecko is a specialist pollinator, it visits many flower species in the Canary Islands.

This error is known as a comma splice. There are three ways of correcting it:

1. by putting a period after *pollinator* and starting a new sentence:

    > ✔ . . . is a specialist pollinator. It . . .

2. by replacing the comma with a semicolon:

   ✔ ... is a specialist pollinator; it ...

3. by making one of the independent clauses subordinate to the other, so that it doesn't stand by itself:

   ✔ The gecko, which is a specialist pollinator, visits many flower species in the Canary Islands.

The one exception to the rule that independent clauses cannot be joined by a comma arises when the clauses are very short and arranged in a tight sequence:

✔ The chemicals mixed, they glowed, they exploded.

You should not use this kind of sentence very often.

Contrary to what many people think, conjunctive adverbs—words such as *however*, *therefore*, and *thus*—cannot be used to join independent clauses:

✘ Two of the sheep developed incurable scrapie, however they were quickly quarantined.

This mistake can be corrected by beginning a new sentence after *scrapie* or (preferably) by replacing the comma with a semicolon:

✔ Two of the sheep developed incurable scrapie; however, they were quickly quarantined.

In this case, a comma is also needed to prevent the reader from misreading *however* as "in whatever way."

Another option is to join the two independent clauses with a coordinating conjunction—*and*, *or*, *nor*, *but*, *for*, *yet*, *so*, or *whereas*:

✔ Two of the sheep developed incurable scrapie, but they were quickly quarantined.

Faulty predication

When the subject of a sentence is not grammatically connected to what follows (the predicate), the result is faulty predication:

✗ The reason the experiment failed was because the pH of the solution was incorrect.

The problem with this sentence is that *the reason* and *was because* mean essentially the same thing. The subject is a noun and the verb *was* needs a noun clause to complete it:

✔ The reason the experiment failed was that the pH of the solution was incorrect.

Another solution is to rephrase the sentence:

✔ The experiment failed because the pH of the solution was incorrect.

Faulty predication also occurs with *is when* and *is where* constructions:

✗ One sign of bear aggression is when the bear swings its head from side to side.

Again, you can correct this error in one of two ways:

1. Follow the *is* with a noun phrase to complete the sentence:

✔ One sign of bear aggression is a side-to-side swinging movement of the bear's head.

(or)

✔ One sign of bear aggression is the bear's swinging of its head from side to side.

2. Change the verb:

✔ One sign of bear aggression occurs when the bear swings its head from side to side.

Subject–Verb Agreement

Identifying the subject

A verb should always agree in number with its subject. Sometimes, however, when the subject does not come at the beginning of the sentence or when it is separated from the verb by other information, you may be tempted to use a verb form that does not agree:

- ✗ The increase in harvested fish and bycatch were condemned by the biologists.

The subject here is *increase*, not *fish and bycatch*; therefore, the verb should be singular:

- ✔ The increase in harvested fish and bycatch was condemned by the biologists.

Either, neither, each

The indefinite pronouns *either*, *neither*, and *each* always take singular verbs:

- ✔ Neither duck species is found in Canada.
- ✔ Each of them is an endangered species.

Compound subjects

When *or*, *either . . . or*, or *neither . . . nor* is used to create a compound subject, the verb should usually agree with the last item in the subject:

- ✔ Neither abscisic acid nor its catabolites were effective repressors.
- ✔ Either the calculations or the graph was incorrect.

You may find, however, that it sounds awkward in some cases to use a singular verb when a singular item follows a plural item:

- **orig.** Either my chemistry books or my biology text is going to gather dust this weekend.

In such instances, it's better to rephrase the sentence:

- **rev.** This weekend, I'm going to ignore either my chemistry books or my biology text.

Unlike the word *and*, which creates a compound subject and therefore takes a plural verb, the phrases *as well as* and *in addition to* do not create compound subjects; therefore the verb remains singular:

- ✔ Spruce and birch are boreal forest trees.
- ✔ Spruce, as well as birch, is a boreal forest tree.

Collective nouns

A collective noun is a singular noun that comprises a number of members, such as *family*, *army*, or *team*. If the noun refers to the members as one unit, it takes a singular verb:

- ✔ The flock is located in a heronry near Tsawwassen.

If, in the context of the sentence, the noun refers to the members as individuals, the verb becomes plural:

- ✔ Juvenile herons in the flock are going to be banded tomorrow.
- ✔ The majority of insect fossils occur in sedimentary layers found just above the riverbed.

Titles

The title of a book or movie or the name of a business or organization is always treated as a singular noun, even if it contains plural words; therefore, it takes a singular verb:

- ✔ *Sharks of the Pacific Northwest* was a thoughtful book.
- ✔ Goodman & Goodman is handling the legal dispute.

Verb Tenses

Native speakers of English usually know without thinking which verb tense to use in a given context. However, a few tenses can still be confusing.

The past perfect

If the main verb is in the past tense and you want to refer to something that happened before that time, use the past perfect (*had* followed by the past participle). The time sequence will not be clear if you use the simple past in both clauses:

- ✗ The wolverine smelled the trap that the hunter placed in the snow.
- ✓ The wolverine smelled the trap that the hunter had placed in the snow.

Similarly, when you are reporting what someone said in the past—that is, when you are using past indirect discourse—you should use the past perfect tense in the clause describing what was said:

- ✗ He told the TA that he wrote the lab report that week.
- ✓ He told the TA that he had written the lab report that week.

Using *if*

When you are describing a possibility in the future, use the present tense in the condition (*if*) clause and the future tense in the consequence clause:

- ✓ If the temperature sinks below −40°C, the bark beetles will die.

When the possibility is unlikely, it is conventional—especially in formal writing—to use the subjunctive in the *if* clause, and *would* followed by the base verb in the consequence clause:

- ✓ If the temperature were to sink below −60°C, the expedition would be cancelled.

When you are describing a hypothetical instance in the past, use the past subjunctive (it has the same form as the past perfect) in the *if* clause and *would have* followed by the past participle for the consequence. A common error is to use *would have* in both clauses:

- ✗ If the stem would have been thornier, I would have used thicker protective gloves.
- ✓ If the stem had been thornier, I would have used thicker protective gloves.

Writing about published work

When you are describing a published work in its historical context, use the past tense:

✔ Richard Dawkins wrote *The Selfish Gene* at a time when many people were questioning evolutionary mechanisms.

To discuss what goes on within a scientific paper, however, you should use the present tense:

✔ The researcher sits in the duck blind and counts the number of female ducks that the decoy attracts.

When you are discussing an episode or incident in a scientific work and want to refer to a prior incident or a future one, use past or future tenses accordingly:

✔ The team returns to northern Quebec, where they spent two summers working at a field station; by the time they leave, they will have measured moss cover at five sites.

Be sure to return to the present tense when you have finished referring to events in the past or future.

Pronouns

Pronoun reference

The link between a pronoun and the noun it refers to must be clear. If the noun doesn't appear in the same sentence as the pronoun, it should appear in the preceding sentence:

✘ The test tube supply in our biology lab had run out, so we borrowed them from the chemistry lab.

Since *test tube* is used as an adjective rather than a noun, it cannot serve as referent or antecedent for the pronoun *them*. You must either replace *them* or change the phrase *test tube supply*:

- ✔ The test tube supply in our biology lab had run out, so we borrowed test tubes from the chemistry lab.
- ✔ Our biology lab had run out of test tubes, so we borrowed them from the chemistry lab.

When a sentence contains more than one noun, make sure there is no ambiguity about which noun the pronoun refers to:

- ✘ The public wants better environmental regulation along with lower taxes, but the government does not favour them.

What does the pronoun *them* refer to: the regulation, the taxes, or both?

- ✔ The public wants better environmental regulation along with lower taxes, but the government does not favour lowering taxes.

Using *it* and *this*

Using *it* and *this* without a clear referent can lead to confusion:

- ✘ Although horizontal gene transfer is common with this strain of bacteria, this took much longer than expected.
- ✔ Although horizontal gene transfer is common with this strain of bacteria, conjugation took much longer than expected.

Make sure that *it* or *this* clearly refers to a specific noun or pronoun.

Using *me* and other objective pronouns

Remembering that it is wrong to say "Jim and me were invited to present our findings to the delegates" rather than "Jim and I were invited ...," many people use the subjective form of the pronoun even when it should be objective:

- ✗ The TA <u>asked</u> Jim and <u>I</u> to submit our results.
- ✓ The TA <u>asked</u> Jim and <u>me</u> to submit our results.

The verb *asked* requires an object; *me* is the objective case. A good way to tell which form is correct is to ask yourself how the sentence would sound with only the pronoun. You will know by ear that the subjective form—"The TA asked I"—is not appropriate.

The same problem often arises with prepositions, which should also be followed by a noun or pronoun in the objective case:

- ✗ <u>Between</u> you and <u>I</u>, this result doesn't make sense.
- ✓ <u>Between</u> you and <u>me</u>, this result doesn't make sense.

- ✗ Eating well is a problem <u>for</u> <u>we</u> students.
- ✓ Eating well is a problem <u>for</u> <u>us</u> students.

There are times, however, when the correct case can sound stiff or awkward:

orig. <u>To</u> <u>whom</u> was the scalpel entrusted?

Rather than using a correct but awkward form, try to reword the sentence:

rev. <u>Who</u> <u>was</u> <u>entrusted</u> with the scalpel?

Exceptions for pronouns following prepositions

The rule that a pronoun following a preposition takes the objective case has exceptions. When the preposition is followed by a clause, the pronoun should take the case required by its position in the clause:

- ✗ The conference organizers showed interest <u>in</u> <u>whom</u> <u>would be selected</u> as the keynote speaker.

Although the pronoun follows the preposition *in*, it is also the subject of the verb *would be selected* and therefore requires the subjective case:

- ✔ The conference organizers showed interest in who would be selected as the keynote speaker.

Similarly, when a gerund (a word that acts partly as a noun and partly as a verb) is the subject of a clause, the pronoun that modifies it takes the possessive case:

- ✘ The inspectors were surprised by him failing to calibrate the mass spectrometer.
- ✔ The inspectors were surprised by his failing to calibrate the mass spectrometer.

Modifiers

Adjectives modify nouns; adverbs modify verbs, adjectives, and other adverbs. Do not use an adjective to modify a verb:

- ✘ The experiment went good. (adjective with verb)
- ✔ The experiment went well. (adverb modifying verb)
- ✔ The experiment went really well. (adverb modifying adverb)

- ✔ The dinosaur fossil from Drumheller had a good cranium. (adjective modifying noun)
- ✔ The dinosaur fossil from Drumheller had a really good cranium. (adverb modifying adjective)

Squinting modifiers

Remember that clarity depends largely on word order: to avoid confusion, the connections between the different parts of a sentence must be clear. Modifiers should therefore be as close as possible to the words they modify. A squinting modifier is one that, because of its position, seems to look in two directions at once:

- ✘ Describing the causes of leishmaniasis clearly helps reduce infection rates.

The meaning of this sentence is ambiguous due to the placement of the modifier *clearly*. Changing the order of the sentence or rephrasing it will make the meaning obvious:

- ✔ Describing the causes of leishmaniasis will clearly help reduce infection rates.
- ✔ A clear description of the causes of leishmaniasis will help reduce infection rates.

Other squinting modifiers can be corrected in the same way:

- ✘ Our biochemistry professor gave a lecture on the three-dimensional structure of proteins, which was well illustrated.
- ✔ Our biochemistry professor gave a well-illustrated lecture on the three-dimensional structure of proteins.

Often the modifier works best when placed immediately in front of the phrase it modifies. Notice the difference that this placement can make:

Only Banting guessed the importance of insulin.
Banting only guessed the importance of insulin.
Banting guessed only the importance of insulin.
Banting guessed the importance of insulin only.

Dangling modifiers

Modifiers that have no grammatical connection with anything else in the sentence are said to be dangling:

- ✘ Walking around the fen in early summer, the grasses and open water made a bleak impression.

Who is doing the walking? Here's another example:

- ✘ Reflecting on the results of the experiment, it was decided not to include the data in the upcoming report.

Who is doing the reflecting? Clarify the meaning by connecting the dangling modifier to a new subject:

- ✔ <u>Walking</u> around the fen in early summer, <u>Darwin</u> thought the trees and open water made a bleak impression.
- ✔ <u>Reflecting</u> on the results of the experiment, <u>we</u> decided not to include the data in the upcoming report.

Pairs and Parallels

Comparisons

Faulty comparisons are perhaps the most common writing mistake in lab reports and essays. Problems with comparisons result from a number of factors, including faulty logic, poor writing, and sloppy revision.

Comparisons must be complete:

- ✘ *Xenopus* egg RNA was less abundant. (Less abundant than what?)
- ✘ Seagrass beds were not as productive. (Not as productive as what?)

The second element in a comparison should be equivalent to the first, whether the equivalence is stated or merely implied:

- ✘ Polls show that Americans have a greater awareness of the environmental impact of pipelines than Canadians.

This sentence suggests that the two things being compared are *pipelines* and *Canadians*. Adding a second verb (*have*) equivalent to the first one shows that the two things being compared are *Americans' awareness* and *Canadians' awareness*:

- ✔ Polls show that Americans <u>have</u> a greater awareness of the environmental impact of pipelines than Canadians <u>have</u>.

A similar problem arises in the following comparison:

- ✘ Mount Temple is <u>a very exposed peak</u> and so are the scree slopes.

The scree slopes may be very exposed, but they are not *a very exposed peak*; to make sense, the two parts of the comparison must be parallel:

✔ The peak of Mount Temple is very exposed and so are the scree slopes.

The following comparison is also incorrect:

✘ The rubber trees in Brazil are superior to Singapore.

The example above incorrectly compares *rubber trees* to *Singapore*. Make sure your comparisons are logical.

✔ The rubber trees in Brazil are superior to the rubber trees in Singapore.
✔ The rubber trees in Brazil are superior to those in Singapore.

Correlatives

Constructions such as *both . . . and, not only . . . but also*, and *neither . . . nor* are especially tricky. For the implied comparison to work, the two parts that come after the coordinating term must be grammatically equivalent:

✘ Dr. Wells not only studies amphibians but also mammals.
✔ Dr. Wells studies not only amphibians but also mammals.

Parallel phrasing

A series of items in a sentence should be phrased in parallel wording. Make sure that all the parts of a parallel construction are in fact equal:

✘ We had to turn in our rough notes, our calculations, and finished assignment.
✔ We had to turn in our rough notes, our calculations, and our finished assignment.

Once you have decided to include the pronoun *our* in the first two elements, the third must have it too.

For clarity as well as stylistic grace, keep similar ideas in similar form:

✗ The oak leaves <u>turned</u> brown and <u>wilted</u>, but abscission did not occur.
✔ The oak leaves <u>turned</u> brown and <u>wilted</u> but <u>did not</u> abscise.

Faulty parallelism is a common problem in bulleted or numbered lists:

✗ There are several reasons for studying this organism:

- large <u>size</u>
- <u>there</u> is a long history of experimentation
- <u>medically</u> important
- <u>detecting</u> it is quick

✔ There are several reasons for studying this organism:

- large <u>size</u>
- long <u>history of experimentation</u>
- medical <u>importance</u>
- easy <u>detection</u>

Summary

Gaining control over written expression is difficult, but essential. Life sciences students frequently complain about marks docked for poor grammar. They reason that grammar is not important, or at least not as important as the science that they are describing. Yet there are a number of good reasons that you should pay attention to grammar. Every fourth-year honours student, every graduate student, and every practising scientist will tell you that without the skill to write up your results in correct English, you will find it difficult to establish a career in this field. No one will make allowances for poor communication. Native English-speaking life scientists already have a huge advantage: since the end of the Second World War, English has become the global language of science. By mastering the grammar of the international language of communication, you will add enormous transportability to your degree. Also, once you improve your grammar, you will find that all essays and reports—no matter the subject—are easier to write.

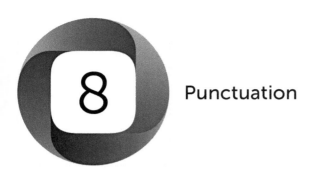

Punctuation

Objective

- Learning the conventions of punctuation: *apostrophe, brackets, colon, comma, dash, ellipsis, exclamation mark, hyphen, italics, parentheses, period, quotation marks,* and *semicolon*

Punctuation causes students so many problems that it deserves a chapter of its own. If your punctuation is faulty, your readers will be confused and may have to backtrack; worse still, they may be tempted to skip over the rough spots. Punctuation marks are the traffic signals of writing; use them with precision to keep readers moving smoothly through your work.

(Items in this chapter are arranged alphabetically.)

Apostrophe [']

1. **Use an apostrophe to indicate possession**. To figure out where to place the apostrophe, think of the possessive as an *of* phrase. If the noun of the *of* phrase ends in *s*, add an apostrophe. If the noun in the *of* phrase does not end in *s*, add an apostrophe plus *s*.

the laboratory of Dr. Jones	→ Dr. Jones' laboratory
the finches of Darwin	→ Darwin's finches
the migration route of the whales	→ the whales' migration route
the panel of the spectrophotometer	→ the spectrophotometer's panel

2. **Use an apostrophe to show contractions of words:**

> isn't; we'll; he's; shouldn't; I'm

Caution: don't confuse *it's* (the contraction of *it is*) with *its* (the possessive of *it*), which has no apostrophe. And remember that possessive pronouns never take an apostrophe: *yours, hers, its, ours, yours, theirs*.

Brackets []

Brackets are square enclosures, not to be confused with parentheses (which are round). **Use brackets to set off a remark of your own within a quotation.** The brackets indicate that the words enclosed are not those of the person quoted:

> Suzuki maintained, "The first step in realizing stronger goals for Canada begins now. Our government promised [in 2016] more ambitious targets and a framework for cutting carbon pollution and expanding renewable energy."

Brackets are sometimes used to enclose *sic*, which is used after an error such as a misspelling to show that the mistake was in the original. *Sic* (literally "*thus, so*") is part of the Latin phrase *sic erat scriptum*, which means "thus was it written." *Sic* may be italicized:

> In his memoirs, Johnson wrote, "It all comes down to heredetary [*sic*] characteristics."

Colon [:]

A colon indicates that something is to follow.

1. **Use a colon before a formal statement or series:**

> ✔ Marine mammals include the following: whales, walruses, dolphins, and seals.

Do not use a colon if the words preceding it do not form a complete sentence:

✗ Marine mammals include: whales, walruses, dolphins, and seals.
✔ Marine mammals include whales, walruses, dolphins, and seals.

On the other hand, a colon often precedes a vertical list, even when the introductory part is not a complete sentence:

✔ Marine mammals include:
whales
walruses
dolphins
seals

2. **Use a colon for formality before a direct quotation or when a complete sentence precedes the quotation:**

   Dobzhansky repeated his message: "Nothing in biology makes sense except in the light of evolution."

3. **Use a colon between numbers expressing time and ratios:**

   4:30 p.m.
   The ratio of calcium to potassium should be 7:1.

Comma [,]

Commas are the trickiest of all punctuation marks; even the experts differ on when to use them. Most agree, however, that too many commas are as bad as too few because they make writing choppy and awkward to read. Certainly recent writers use fewer commas than earlier stylists did. Whenever you are in doubt, let clarity be your guide. The most widely accepted conventions are these:

1. **Use a comma to separate two independent clauses joined by a coordinating conjunction (*and, but, for, or, nor, yet, so, whereas*).** By signalling that there are two clauses, the comma will prevent the reader from thinking that the beginning of the second clause is the end of the first:

   ✗ He sampled the fish with help of his lab partner and the technician ran the analysis.

✔ He sampled the fish with help of his lab partner, and the technician ran the analysis.

When the second clause has the same subject as the first, you have the option of omitting both the second subject and the comma:

✔ Gazelles can run quickly, but they cannot run any great distance.
✔ Gazelles can run quickly but cannot run any great distance.

If you mistakenly punctuate two sentences as if they were one, the result will be a run-on sentence; if you use a comma but forget the coordinating conjunction, the result will be a comma splice:

✘ We wanted to study the bison herd, it was under quarantine.
✔ We wanted to study the bison herd, but it was under quarantine.

Remember that words such as *however*, *therefore*, and *thus* are conjunctive adverbs, not conjunctions; if you use one of them to join two independent clauses, the result will again be a comma splice:

✘ We measured the trees after a year, however, we waited two years to collect the cones.
✔ We measured the trees after a year; however, we waited two years to collect the cones.

Conjunctive adverbs are often confused with conjunctions. You can distinguish between the two if you remember that a conjunctive adverb's position in a sentence can be changed:

✔ We measured the trees after a year; we waited two years, however, to collect the cones.

The position of a conjunction, on the other hand, is invariable; it must be placed between the two clauses:

✔ We measured the trees after a year, but we waited two years to collect the cones.

A good rule of thumb, then, is to use a comma when the linking word can't move.

When, in rare cases, the independent clauses are short and closely related, they may be joined by a comma alone:

✔ Crows caw, ducks quack, geese honk.

2. **Use a comma between items in a series.** Place a coordinating conjunction before the last item:

   ✔ The eagle ate a female snake that was large, healthy, and gravid.
   ✔ The equipment included an HPLC, a centrifuge, three 96-well plates, and various solvents.

   The comma before the conjunction is optional for single items in a series:

   ✔ The arboretum has Douglas-fir, sequoia and Wollemi pine.

   For phrases in a series, however, use the final comma to help to prevent confusion:

   ✗ Before going on the Juan de Fuca Marine Trail, we were warned about ticks, attacks by bears and high tides.

   In this case, a comma would prevent the reader from thinking that attacks were made by bears as well as high tides:

   ✔ We were warned about ticks, attacks by bears, and high tides.

3. **Use a comma to separate adjectives preceding a noun when they modify the same element:**

   ✔ Golden hamsters live in the open, windswept steppe.

   However, when the adjectives do not modify the same element, you should not use a comma:

   ✗ It was a successful, winter field trip.

Here *winter* modifies *field trip*, but *successful* modifies the whole phrase—*winter field trip*. A good way of deciding whether or not you need a comma is to see if you can reverse the order of the adjectives. If you can reverse them (*open, windswept steppe* or *windswept, open steppe*), use a comma; if you can't (*winter successful field trip*), omit the comma:

- ✔ It was a successful winter field trip.

4. **Use commas to set off an interruption (or "parenthetical element"):**

    - ✔ That particular lek site, Anholt realized, was mistakenly included in the experiment.
    - ✔ The data, however, did not support the hypothesis.

    Remember to put commas on both sides of the interruption:

    - ✘ The data however, did not support the hypothesis.
    - ✘ The software, they claim is not well supported by the company.
    - ✔ The software, they claim, is not well supported by the company.

5. **Use commas to set off words or phrases that provide additional but non-essential information:**

    - ✔ The dominant tree, *Larix occidentalis*, is host to many insects.
    - ✔ The golden retriever, the imprinted duck's new companion, went with him everywhere.

    In these examples, *Larix occidentalis* and *the imprinted duck's new companion* are *appositives*: they give additional information about the nouns they refer to (*tree* and *golden retriever*), but the sentences would make sense without them. Here's another example:

    - ✔ Krummholz tree forms, which used to be found in Cape Breton, are increasingly rare.

    The phrase *which used to be found in Cape Breton* is a non-restrictive modifier because it does not limit the meaning of the term it modifies (*krummholz tree forms*). Without that modifying clause, the sentence would still specify what forms are increasingly rare. Because the information the clause provides is not necessary to the meaning of the sentence, you must use commas on both sides to set it off.

In contrast, a restrictive modifier is one that provides essential information; it must not be set apart from the element it modifies, and commas should not be used:

✔ The woman who is wearing the green lab coat is my lab partner.

The phrase *who is wearing the green lab coat* cannot be detached without the sentence losing its meaning.

To avoid confusion, be sure to distinguish carefully between essential and additional information. The difference can be important:

Humans, who test HIV-positive, have shorter life expectancies.
(All humans test HIV-positive and have shorter life expectancies.)

Humans who test HIV-positive have shorter life expectancies.
(Only humans who test HIV-positive have shorter life expectancies.)

6. **Use a comma after an introductory phrase when omitting it would cause confusion:**

    ✘ On the escarpment above the trees were over 800 years old.
    ✔ On the escarpment above, the trees were over 800 years old.

    ✘ When the wolf turned away the moose escaped.
    ✔ When the wolf turned away, the moose escaped.

7. **Use a comma to separate elements in dates and addresses:**

    ✔ February 2, 2015 (Commas are often omitted if the day comes first: 2 February 2015.)
    ✔ 117 Hudson Drive, Edmonton, Alberta
    ✔ They lived in Dartmouth, Nova Scotia.

8. **Use a comma before a quotation in a sentence:**

    ✔ Steven Pinker wrote, "Just as blueprints don't necessarily specify blue buildings, selfish genes don't necessarily specify selfish organisms."
    ✔ "Enrolling in the diving certification program," the director told us, "is mandatory."

    For more formality, or if the quotation is preceded by a complete sentence, you may use a colon (see page 142).

9. **Use a comma with a name followed by a title:**

    ✔ John Dutcher, President
    ✔ Gilles Lavoie, Ph.D.

10. **Do not use a comma between a subject and its verb:**

    ✗ The most common ligand of all, is vanillin.
    ✔ The most common ligand of all is vanillin.

11. **Do not use a comma between a verb and its object:**

    ✗ The rat in the maze decided, what it must do.
    ✔ The rat in the maze decided what it must do.

12. **Do not use a comma between a coordinating conjunction and the following clause:**

    ✗ Clams were selected at all time points but, mussels were selected only once.
    ✔ Clams were selected at all time points, but mussels were selected only once.

Dash [—]

A dash creates an abrupt pause, emphasizing the words that follow. Never use dashes as casual substitutes for other punctuation; overuse can detract from the calm, well-reasoned effect you want to create. Most often, you will use a dash to stress a word or a phrase:

> This year 68,000 harp seals were killed—nearly twice the number of seals that were killed last year.

> Leaf caterpillars were offered a diet of two species of poplars—the insects refused them both.

You can type two hyphens together, with no spaces on either side, to show a dash; your word processor may automatically convert this to a solid line as you continue typing. Alternatively, you can insert an em-dash from the list of special characters in your word-processing program.

Ellipsis [. . .]

1. **Use an ellipsis (three dots) to show an omission from a quotation:**

   For an ellipsis within a sentence, use three periods with a space before each and a space after the last:

   > "The animal care committee reported that post-mortem tests . . . verified that poor nutrition was a factor in the deaths of the mice."

   If the omission comes at the beginning of the quotation, an ellipsis is not necessary:

   > The committee cited evidence that "verified that poor nutrition was a factor in the deaths of the mice."

   When the omission comes at the end of a sentence, use four periods with no space before the first or after the last:

   > The committee noted that "poor nutrition was a factor. . . ."

2. **Use an ellipsis to show that a series of numbers continues indefinitely:**

   > 1, 3, 5, 7, 9 . . .

Exclamation Mark [!]

An exclamation mark helps to show emotion or feeling. It is usually found in dialogue:

> "Eureka!" he cried.

In academic writing, you should use it only in those rare cases when you want to give a point emotional emphasis:

> The model had predicted that the glacier would disappear in five decades. It was gone in two!

Hyphen [-]

1. **Use a hyphen if you must divide a word at the end of a line.** Although it's generally best to start a new line if a word is too long, there are instances—for example, when you're formatting text in narrow columns—when hyphenation might be preferred. The hyphenation feature in current word-processing programs has taken the guesswork out of dividing words at the end of the line, but in the event that you use manual hyphenation, here are a few guidelines:

    - Divide between syllables.
    - Never divide a one-syllable word.
    - Never leave one letter by itself.
    - Divide double consonants except when they come before a suffix, in which case divide before the suffix:

        ar-rangement
        success-ful
        fall-ing
        pass-able

    - When the second consonant has been added to form the suffix, keep it with the suffix:

        drop-ping
        begin-ning

2. **Use a hyphen to separate the parts of certain compound words:**

    - compound nouns:

        close-up; mother-of-pearl

    - compound verbs:

        field-test; dive-bomb

    - compound modifiers:

        a time-consuming experiment; a well-researched paper

    Note that compound modifiers are hyphenated only when they precede the part modified; otherwise, omit the hyphen:

    The experiment was time consuming.
    His paper was well researched.

    Also, do not hyphenate a compound modifier that includes an adverb ending in *-ly*:

    ✔ a well-written paper
    ✘ a beautifully-written paper
    ✔ a beautifully written paper

    Spell-check features today will help you determine which compounds to hyphenate, but there is no clear consensus even from one dictionary to another. As always, consistency in your writing style is most important.

3. **Use a hyphen with certain prefixes (*all-*, *self-*, *ex-*) and with prefixes preceding a proper name.** Again, practices vary, so when in doubt consult a dictionary.

    all-Canadian; self-pollination; ex-biologist

4. **Use a hyphen to emphasize contrasting prefixes:**

    It was necessary to control for pre- and post-treatment damage to the plants.

5. **Use a hyphen to separate written-out compound numbers from one to ninety-nine, and compound fractions:**

    eighty-one years ago; seven-tenths full; two-thirds of the herd

6. **Use a hyphen to separate parts of inclusive numbers or dates:**

    the years 1890-1914; pages 3-10

Italics [*italics*]

1. **Use italics for the titles of works published independently, such as books, long poems that are complete books, plays, films, CDs, and long musical compositions:**

    Birdscapes by Jeremy Mynott has been described as a rare philosophical exploration of our relationship with birds.

    For articles, essays, short poems, or songs, use quotation marks.

2. **Use italics to emphasize an idea:**

    It is important that all glassware be washed *immediately*.

    Be sparing with this use, interspersing it with other, less intrusive methods of creating emphasis.

3. **Use italics (or quotation marks) to identify a word or phrase that is itself the subject of discussion:**

    In biology, the phrase *Red Queen Hypothesis* refers to the evolutionary arms race between prey and predator.

4. **Use italics for foreign words or expressions that have not been naturalized in English:**

> Goethe summed up floral structure with the phrase *Alles ist Blatt*—"all is leaf."

> The *raison d'être* for the conservation of this small RNA is its key role in stress responses.

Parentheses [( )]

1. **Use parentheses to enclose an explanation, example, or qualification.** Parentheses show that the enclosed material is of incidental importance to the main idea. They make an interruption that is more subtle than one marked off by dashes but more pronounced than one set off by commas:

> Ferns have 160 Gb DNA/cell (the highest record for plants), which is much less than that of amoebae.

> Their latest plan (according to everyone) is to log that stretch of forest.

Remember that punctuation should not precede parentheses but may follow them if required by the sense of the sentence:

> Lancetfish prefer to eat squid rather than salps (if given a choice), but they are known to engage in cannibalism, too.

If the parenthetical statement comes between two complete sentences, it should be punctuated as a sentence, with the period, question mark, or exclamation mark inside the parentheses:

> We found many red trilliums (*Trillium erectum*) in Point Pelee National Park. (They were common in woods and thickets.) A rare find was the drooping trillium (*Trillium flexipes*).

2. **Use parentheses to enclose references.** See Chapter 11 for details.

Period [.]

1. **Use a period at the end of a sentence.** A period indicates a full stop, not just a pause.

2. **Use a period with some abbreviations.** It is still common, although not mandatory, to use periods in abbreviated titles (Mrs., Dr., Rev., etc.), academic degrees (M.S.W., Ph.D., etc.), and expressions of time (6:30 p.m.).

    However, Canada's adoption of the metric system in 1970 contributed to a trend away from the use of periods in many abbreviations. State and provincial abbreviations do not require periods (BC, NT, PE, NY, DC). In addition, most acronyms for organizations do not use periods (CIDA, CBC, UNESCO, NSERC).

3. **Use a period at the end of an indirect question.** Do not use a question mark:

    ✗ The TA asked if I wanted to try this method on another bacterium?
    ✓ The TA asked if I wanted to try this method on another bacterium.

    ✗ Scientists from Fisheries and Oceans Canada wondered where the coho salmon went?
    ✓ Scientists from Fisheries and Oceans Canada wondered where the coho salmon went.

4. **Use a period for questions that are really polite orders:**

    Will you please send him the analysis by Friday.

Quotation Marks [" "]

1. **Use quotation marks to signify direct discourse (the actual words of a speaker):**

    I asked, "What is the formula?"
    "I left the gel rig in the top drawer," she replied.

2. **Use quotation marks to show that words themselves are the issue:**

   The prefix "arrheno-" comes from the Greek word for "male."

Alternatively, you may italicize the terms in question.
   Sometimes quotation marks are used to mark a slang word or inappropriate usage to show that the writer is aware of the difficulty:

   Several of the local whale experts did not seem to know about "rogue male" behaviour.

   Use this device only when necessary. In general, it's better to let the context show your attitude or to choose another term.

Semicolon [;]

1. **Use a semicolon to join independent clauses (complete sentences) that are closely related:**

   - ✔ For five days they ran transects; by Saturday the team was exhausted.
   - ✔ The herons were panicking; they could not defend their nests from the attacking eagle.

   A semicolon is especially useful when the second independent clause begins with a conjunctive adverb such as *however, moreover, consequently, nevertheless, in addition,* or *therefore* (usually followed by a comma):

   - ✔ Thaumatin-like protein was isolated in nanomolar quantities; consequently, we were able to proceed with mass spectrometry.

   It's usually acceptable to follow a semicolon with a coordinating conjunction if the second clause is complicated by other commas:

   - ✔ Penguins move about in all weather; but sometimes, especially in the Antarctic winter, movement can be fatal.

2. **Use a semicolon to mark the divisions in a complicated series when individual items themselves need commas.** Using a comma to mark the subdivisions and a semicolon to mark the main divisions will help to prevent mix-ups:

   ✗ We selected genotypes S125, the highest-yielding one, S4, and S71.

   Is the highest-yielding one S125, S4, or a separate genotype?

   ✔ We selected genotype S125; the highest-yielding one, S4; and S71.

   In a case such as this, the elements separated by the semicolon need not be independent clauses.

Summary

Punctuation problems trouble many students. Some are so afraid of misusing punctuation that they write exclusively in short, sparsely punctuated sentences. Yet to explore complex scientific topics in detail, writers must use longer, carefully constructed sentences. Not having the option to follow the ancient Roman practice of using no punctuation marks whatsoever, you need to understand and follow the basic rules. When used correctly, punctuation will help you control the dynamics of your writing and guide your reader through your ideas. From the obvious emphasis provided by exclamation marks, to pauses provided by commas, to asides provided by parentheses, punctuation is powerful. By learning the rules of punctuation, you will gain the confidence to create sentences and phrases that are not only direct and unambiguous, but that have an engaging flow and rhythm.

9 Misused Words and Phrases

Objectives

- Differentiating between words that sound similar but have different meanings
- Avoiding common spelling mistakes
- Choosing the word that suits the context
- Recognizing commonly confused singular and plural nouns

Here are some words and phrases that are often misused. If you're wondering about a particular word or idiom, check here for advice about correct usage.

accept, except. *Accept* is a verb meaning "to receive affirmatively"; *except*, when used as a verb, means "to exclude":

> Our research group <u>accepted</u> applicants with experience in tissue culture.
> The researcher <u>excepted</u> unhealthy individuals from the general trial.

accompanied by, accompanied with. Use *accompanied by* for people and animals; use *accompanied with* for objects:

> In early spring, the female caribou were <u>accompanied by</u> last year's calves.
> The sledge microtome arrived, <u>accompanied with</u> an instruction manual.

advice, advise. *Advice* is a noun, *advise* a verb:

> We were <u>advised</u> to ignore the <u>advice</u> published online.

affect, effect. *Affect* is a verb meaning "to influence"; however, it also has a specialized meaning in psychology, referring to a person's emotional state. *Effect* can be either a noun meaning "result" or a verb meaning "to bring about":

> Uninterrupted work at a computer affects vision.
>
> Because he was so depressed, he showed no affect when the bell was rung.
>
> The effect of long exposure to ultraviolet light is DNA damage.
>
> High temperatures can effect intense chemical reactions.

all ready, already. To be *all ready* is simply to be ready for something; *already* means "beforehand" or "earlier":

> The monarch butterflies were all ready to begin their migration.
>
> By the time the coyotes arrived in Montgomery Woods, the deer had already left.

all right. Write as two separate words: *all right*. This can mean "safe and sound, in good condition, okay"; "correct"; "satisfactory"; or "I agree":

> Are you all right?
> The student's answers were all right.

(Note the ambiguity of the second example: does it mean that the answers were all correct or simply satisfactory? In this case, it might be better to use a clearer word.)

all together, altogether. *All together* means "in a group"; *altogether* is an adverb meaning "entirely":

> The last time so many orcas were all together was in 2017.
>
> The results were altogether confusing.

allusion, illusion. An *allusion* is an indirect reference to something; an *illusion* is a false perception:

> The comment about bills is an allusion to Darwin's original study of finches.

Heat over sand can cause an inferior mirage, which is an optical <u>illusion</u>.

a lot. Write as two separate words: *a lot*.

alternate, alternative. *Alternate* means "every other or every second thing in a series"; *alternative* refers to a choice between options:

Oocyte development was measured on <u>alternate</u> days.

For protein separation, an <u>alternative</u> to gel electrophoresis is liquid–liquid extraction.

among, between. Use *among* for three or more persons or objects, *between* for two:

The pigs root <u>among</u> the oaks.

There is a Garry oak meadow <u>between</u> Mount Tolmie and Mount Douglas.

amount, number. *Amount* indicates quantity when units are not discrete and not absolute; *number* indicates quantity when units are discrete and absolute:

There has been a large <u>amount</u> of forest loss due to bark beetle damage.

A large <u>number</u> of birds were observed at Montmagny.

See also **less, fewer**.

analysis. The plural is *analyses*.

anyone, any one. *Anyone* is written as two words to give numerical emphasis; otherwise it is written as one word:

<u>Any one</u> of the researchers can access the database at any time.

As <u>anyone</u> who has studied in the Canadian Arctic knows, freezing temperatures can occur even on summer nights.

anyways. Non-standard. Use *anyway*.

as, because. *As* is a weaker conjunction than *because* and may be confused with *when*:

- ✗ As it was winter, the animals exhibited slow response times.
- ✓ Because it was winter, the animals exhibited slow response times.

as to. A common feature of bureaucratese. Replace it with a single-word preposition such as *about* or *on*:

- ✗ Ornithologists were uncertain as to the distribution of the ivory-billed woodpecker.
- ✓ Ornithologists were uncertain about the distribution of the ivory-billed woodpecker.

bad, badly. *Bad* is an adjective meaning "not good":

> The autumn weather in Prince Rupert is bad.

Badly is an adverb meaning "not well"; when used with the verbs *want* or *need*, it means "very much":

> We discarded one of the cytokinins because it was badly manufactured and contained numerous impurities.

> A solution to this public health problem is badly needed.

beside, besides. *Beside* is a preposition meaning "next to":

> In this region, large stands of poplar are commonly found beside streams and rivers.

Besides has two uses: as a preposition it means "in addition to"; as a conjunctive adverb it means "moreover":

> Besides studying poplar, we are also examining alder and willow.

> We abandoned the first method because it was laborious; besides, one of the steps involved cyanide chemistry, which raised health and safety issues.

between. See **among**.

bring, take. One *brings* something to a closer place and *takes* it to a farther one:

> To complete the analysis, students must bring the plants from the greenhouse into the laboratory.
>
> The advance team took the diatom sampler to the outpost.

can, may. *Can* means "to be able"; *may* means "to have permission":

> Can this theory be tested?
>
> May I have another chance to test this hypothesis?

In speech, *can* is used to cover both meanings; in formal writing, however, you should observe the distinction.

can't hardly. A faulty combination of the phrases *can't* and *can hardly*. Use one or the other:

> Some breeds of dog can hardly swim.

cite, sight, site. To *cite* something is to quote or mention it as an example or authority; *sight* can be used in many ways, all of which relate to the ability to see; *site* refers to a specific location, a particular place at which something is located:

> As technology advanced, these early studies were no longer cited.
>
> A vole's sight is extremely limited.
>
> Although the murrelet returned annually to the same site, it always nested in a new burrow.

complement, compliment. The verb *to complement* means "to complete or enhance"; *to compliment* means "to praise":

> The analysis of pollen types in cores taken from lake bottoms complements the dendrochronology study from the previous year.
>
> The agency official complimented Dr. Kong for his outstanding report.

compose, comprise. Both words mean "to constitute or make up," but *compose* is preferred. *Comprise* is correctly used to mean "include," "consist of," or "be composed of." Using *comprise* in the passive ("is comprised of")—as you might be tempted to do in the second example below—is usually frowned on in formal writing:

> Bone is composed of proteins and minerals.
>
> Each kit comprises a number of different enzymes.

continual, continuous. *Continual* means "repeated over a period of time"; *continuous* means "constant" or "without interruption":

> The unseasonably heavy rain caused continual delays in building the observation blind.
> After five days of continuous rain, the vernal ponds appeared.

could of. This construction is incorrect, as are *might of, should of,* and *would of.* Replace *of* with *have*:

- ✗ Better results could of been achieved if we had not been so rushed.
- ✓ Better results could have been achieved if we had not been so rushed.

- ✓ Another test might have been possible, but the early frosts this year precluded further work.
- ✓ Safety officers should have known about the impending storm and warned collection crews.
- ✓ The bears would have left earlier, but they stayed longer because of the abundant fruit.

council, counsel. *Council* is a noun meaning "an advisory or deliberative assembly." *Counsel* as a noun means "advice" or "lawyer"; as a verb it means "to give advice."

> An advisory council should be formed to generate appropriate floodplain management policy for the Ministry of Forests.

> They acknowledged the timely counsel of Natalie Prince.
> He counsels first-year students.

criterion, criteria. A *criterion* is a standard for judging something. *Criteria* is the plural of *criterion* and thus requires a plural verb:

There are five criteria for grading the degree of insect damage.
The major criterion was excellence of experimental design.

data. The plural of *datum*. The set of information, usually in numerical form, that is used for analysis as the basis for a study. Because *data* often refers to a single mass entity, many writers now accept its use with a singular verb and pronoun:

These data were gathered in an unsystematic fashion.
The data is expected to be submitted to the committee by early November.

deduce, deduct. To *deduce* something is to work it out by reasoning; to *deduct* means "to subtract or take away from something." The noun form of both words is *deduction*.

In a recent review, Baum et al. (2012) deduce that the recovery of shark populations is unlikely.
We had to deduct the cost of materials from the research budget.

defence, defense. Both spellings are correct: *defence* is standard in Britain and is somewhat more common in Canada; *defense* is standard in the United States.

dependent, dependant. *Dependent* is an adjective meaning "contingent on" or "subject to"; it also refers to the variable being tested in an experiment (i.e., the dependent variable).

Nectary development is dependent on two transcription factors: BOP1 and BOP2.
In this experiment, heartbeat is the dependent variable, and stress is the independent variable.

Dependant is a noun that refers to someone who relies on someone else for support:

She is a dependant of her mother.

device, devise. The word ending in *-ice* is the noun; the word ending in *-ise* is the verb.

different than, different from. Use *different from* to compare two persons or things; use *different than* with a full clause:

> Swainson's thrushes are quite different from other thrushes.
> This landscape is different than it used to be.

diminish, minimize. *To diminish* means "to make or become smaller"; *to minimize* means "to reduce something to the smallest possible amount or size":

> As mass spectrometers improve in accuracy and precision, the minimum sample volume diminishes.
>
> The new omnibus bill will minimize protection of the nation's rivers and lakes.

disinterested, uninterested. *Disinterested* implies impartiality or neutrality; *uninterested* implies a lack of interest:

> As a disinterested observer, he was in a good position to judge the issue fairly.
>
> Uninterested in the seminar, he yawned repeatedly.

due to. Although increasingly used to mean "because of," *due* is an adjective and therefore needs to modify something:

> ✗ Due to his impatience, we lost the contract. (*Due* is dangling.)
> ✓ The loss was due to his impatience.

e.g., i.e. *E.g.* means "for example"; *i.e.* means "that is." It is incorrect to use them interchangeably.

entomology, etymology. *Entomology* is the study of insects; *etymology* is the study of the derivation and history of words.

exceptional, exceptionable. *Exceptional* means "unusual" or "outstanding," whereas *exceptionable* means "open to objection" and is generally used in negative contexts:

> David Schindler's accomplishments are exceptional.

Data management of fisheries records would have been deemed <u>exceptionable</u> had it not been for recent implementation of the recommended oversight procedures.

farther, further. *Farther* refers to distance, *further* to extent:

> The coots swam <u>farther</u> along the river than the ducks did.
>
> <u>Further</u> experiments focused on the optical properties of tiger shrimp.

focus. The plural of the noun may be either *focuses* (also spelled *focusses*) or *foci*.

good, well. *Good* is an adjective that modifies a noun; *well* is an adverb that modifies a verb:

> Rice breeders at the international institute have developed a <u>good</u> resistance program.
>
> Green fluorescent proteins work <u>well</u> in this system only when linked to the first of the four promoters.

hanged, hung. *Hanged* means "executed by hanging." *Hung* means "suspended" or "clung to":

> In medieval trials a pig could be sentenced to be <u>hanged</u> for eating crops.
>
> He <u>hung</u> the plants in the drying shed until they reached 20 per cent relative water content.
>
> The lamprey <u>hung</u> on to the side of the fish.

hereditary, heredity. *Heredity* is a noun; *hereditary* is an adjective. *Heredity* is the biological process whereby characteristics are passed from one generation to the next; *hereditary* describes those characteristics:

> <u>Heredity</u> is a factor in the incidence of this disease.
> Your asthma may be <u>hereditary</u>.

i.e. This is not the same as *e.g.* See **e.g.**

incite, insight. *Incite* is a verb meaning "to stir up"; *insight* is a noun meaning "(often sudden) understanding":

> A number of pathways can incite enzyme activation.
>
> Neuroscience provides insight into the well-developed emotions of orcas.

infer, imply. To *infer* means "to deduce or conclude by reasoning." It is often confused with *imply*, which means "to suggest or insinuate":

> We infer from the departure from a 50:50 sex ratio that an epigenetic mechanism is at work.
>
> The phylogenetic position of this species implies an ancestral link to modern turtles.

inflammable, flammable, non-flammable. Despite its *in-* prefix, *inflammable* is not the opposite of *flammable*: both words describe things that are easily set on fire. The opposite of flammable is *non-flammable*. To prevent any possibility of confusion, it's best to avoid *inflammable* altogether.

irregardless. Non-standard. Use *regardless*.

its, it's. *Its* is a form of possessive pronoun; *it's* is a contraction of *it is*. Many people mistakenly put an apostrophe in *its* in order to show possession:

> ✗ The lost cub was looking for it's mother.
> ✔ The lost cub was looking for its mother.
> ✔ It's an unforeseen outcome of this experiment.

less, fewer. *Less* is used when units are not discrete and not absolute (as in "less information"). *Fewer* is used when the units are discrete and absolute (as in "fewer details").

lie, lay. To *lie* means "to assume a horizontal position"; *to lay* means "to put down." The changes of tense often cause confusion:

Present	Past	Past participle	Present participle
lie	lie	lain	lying
lay	laid	laid	laying

- ✗ The wolf was laying in its den.
- ✓ The wolf was lying in its den.
- ✓ We laid the fossil bone fragments in their expected positions.
- ✓ The bison lies down and ruminates.
- ✓ The crew from Fisheries and Oceans Canada was laying the gill net out to dry.

like, as. *Like* is a preposition, but it is often wrongly used as a conjunction. To join two independent clauses, use the conjunction *as*:

- ✗ I want to improve my lab report like you have improved yours.
- ✓ I want to improve my lab report as you have improved yours.
- ✓ Asters are like chrysanthemums.

might of. Incorrect. See **could of.**

minimize. See **diminish.**

mitigate, militate. *To mitigate* means "to reduce the severity of something"; *to militate against* something means "to oppose" it:

> Sewage treatment mitigates river pollution.
>
> The conflict among conservationists militates against the creation of a unified proposal.

myself, me. *Myself* is an intensifier of, not a substitute for, *I* or *me*:

- ✗ The experiment was run by John and myself.
- ✓ The experiment was run by John and me.
- ✗ Jane and myself are responsible for running the gels.
- ✓ Jane and I are responsible for running the gels.
- ✓ I hesitate to mention myself here.

nor, or. Use *nor* with *neither*; use *or* by itself or with *either*:

> The leafcutter bee neither lives in groups nor stores honey.
> The plant is either diseased or dried out.

off of. Remove the unnecessary *of*:

> ✗ The fence kept the deer off of the tree nursery's grounds.
> ✔ The fence kept the deer off the tree nursery's grounds.

phenomenon. A singular noun; the plural is *phenomena*.

populace, populous. *Populace* is a noun meaning "the people of a place"; *populous* is an adjective meaning "thickly inhabited":

> The populace of Manitoulin Island is largely rural.
>
> With so many people in such a small area, Hilltop Village is a populous place.

practice, practise. Both of these spellings have become acceptable for either the noun or the verb. Just be consistent in whatever form you choose.

precede, proceed. To *precede* is to go before (earlier) or in front of others; to *proceed* is to go on or ahead:

> Careful soil preparation in the spring precedes the planting of shrubs and small trees.
>
> Migrating elk proceed across the plain; wolves are not far behind.

prescribe, proscribe. These words are sometimes confused, although they have quite different meanings. *Prescribe* means "to advise the use of" or "to impose authoritatively." *Proscribe* means "to reject, denounce, or ban":

> Dr. Averril prescribed standardized conditions for extracting DNA from mussels.
>
> Municipalities proscribe the feeding of wildlife in urban areas.

principle, principal. *Principle* is a noun meaning "a general truth or law"; *principal* can be used as either a noun, referring to the head of a school or a capital sum of money, or an adjective, meaning "chief":

> The human ethics guidelines are based on the principle that consent should always be sought from participants.

> Daniel Woolf is the Principal of Queen's University.
>
> The principal reason for adding calcium to the medium is that it induces germination of pollen.

promote, inhibit. In the life sciences, the opposite of *promote* is *inhibit*:

> An analog of indoleacetic acid promotes growth of crown-gall tumour tissue.
>
> Heat stress inhibited skeletal muscle hypertrophy.

rational, rationale. *Rational* is an adjective meaning "logical" or "able to reason." *Rationale* is a noun meaning "explanation":

> The controls that were proposed resulted in a rational experimental design.
>
> The rationale behind Dr. Pegworth's thesis was to investigate premature cancer in stressed mice.

real, really. *Real*, an adjective, means "true" or "genuine"; *really*, an adverb, means "actually," "truly," "very," or "extremely":

> The appearance of vesicles was real, and not, as previously suggested, an artifact of preparation.
>
> Populations of bittersweet, *Celastrus scandens*, in Manitoba are really rare further west of Riding Mountain Park.

seasonable, seasonal. *Seasonable* means "usual or suitable for the season"; *seasonal* means "of, depending on, or varying with the season":

> This summer, the temperatures have been quite seasonable.
>
> Phenotype can be affected by seasonal temperature changes.

should of. Incorrect. See **could of**.

that, which. *That* introduces restrictive information that completes the meaning of a clause or sentence; *which* preceded by a comma introduces non-restrictive information that is not essential to the meaning of the clause or sentence:

> The electrophoresis rig that is on the bench needs to be grounded.
> The electrophoresis rig, which is on the bench, needs to be grounded.

In the first sentence, there is more than one rig, but the one requiring grounding is on the bench. In the second sentence, there is only one rig; it is on the bench and needs grounding.

their, there, they're. *Their* is the possessive form of the third person plural pronoun. *There* is usually an adverb, meaning "at that place" or "at that point". *They're* is a contraction of "they are":

> The instructions were to place their seedlings there and nowhere else.
>
> There were no further opportunities to count snow geese.
>
> They're going to migrate along the Pacific Flyway.

tortuous, torturous. The adjective *tortuous* means "full of twists and turns" or "circuitous." *Torturous*, derived from *torture*, means "involving torture" or "excruciating":

> To avoid the female bear and her cub, we took a tortuous route home.
>
> The midnight sampling trip to the seashore was a torturous experience for the students.

translucent, transparent. A *translucent* substance permits light to pass through, but not enough for a person to see through it; a *transparent* substance permits light to pass unobstructed, so that objects can be seen clearly through it.

turbid, turgid. *Turbid*, with respect to a liquid or colour, means "muddy," "not clear," or (with respect to literary style) "confused." *Turgid* means "swollen," "inflated," "enlarged," or (again with reference to literary style) "pompous" or "bombastic."

unique. This word, which means "of which there is only one" or "unequalled," is both overused and misused. Because there are no degrees of comparison—one thing cannot be "more unique" than another—expressions such as *very unique* or *quite unique* are incorrect.

while. To avoid misreading, use *while* only when you mean "at the same time that." Do not use *while* as a substitute for *although*, *whereas*, or *but*:

> ✗   While the results show a trend, they are not significant.

- ✗ I went to collect samples from the trees, while the others decided to stay back at camp.
- ✓ Dogs sometimes make little barking noises while they are sleeping.

who. *Who* can be used to introduce a restrictive clause. *Who* combined with a comma can introduce a non-restrictive clause:

> The nurses gave the vaccine to expectant mothers who were most susceptible to H1N1 virus.
> The nurses gave the vaccine to expectant mothers, who were most susceptible to H1N1 virus.

In the first sentence, not every expectant mother was highly susceptible to H1N1; the nurses vaccinated only those who were highly susceptible. In the second sentence, the nurses vaccinated all expectant mothers because they were all highly susceptible.

-wise. Never use *-wise* as a suffix to form new words when you mean "with regard to":

- ✗ Research-wise, the company did better last year.
- ✓ With regard to research, the company did better last year.

your, you're. *Your* is a possessive adjective; *you're* is a contraction of *you are*:

> Be sure to take your lab book with you.
> You're likely to miss your interview.

Summary

English is full of words that sound similar but have radically different meanings. Many peculiarities within the language developed over time as English evolved without strict regulations or governance. Spelling in English was fluid until the mid-eighteenth century, and this variation is still reflected in some words (e.g., *flammable* and *inflammable*). Today, English continues to evolve, with influences from around the globe. Because scientific vocabulary is a mix of traditional terms and new technical terms, it is essential to master standard usage to achieve clarity in scientific writing.

10 Using Illustrations

Objectives

- Understanding the importance of graphs—data interpretation and display
- Knowing when to use tables
- Presenting frequency data—bar graphs and histograms
- Plotting longitudinal data—line graphs
- Exploring the significance of data—box plots and scatter plots
- Describing data presented in graphs or figures
- Interpreting and avoiding pie charts
- Adding relevant photographs
- Labelling graphs and figures
- Choosing between tables and graphs

In the life sciences, illustrations are not only useful tools for summarizing complex information in reports but also essential components of data interpretation. You should always begin your analysis by graphing your data. Graphs will help you distill the data and assess what the numbers might mean. Just looking at the data won't do—it takes a good graph or table to reveal relationships that are difficult to detect. Graphic representations allow you to analyze your data in sophisticated ways.

Illustrations can also clarify complicated systems. For examples, look at the diagrams you see every day in your textbooks. You can use similar illustrations in your own reports and papers. You might choose to create a model of electron transfer within a membrane, a diagram of the nitrogen cycles in terrestrial ecosystems, or a representation of the lock-and-key model of enzyme function—each is an example of graphic simplification. Illustrations often provide a far more effective summary than you could achieve in writing. A model can be worth a thousand words. (See Chapter 12 for more information on how you can use diagrams to enhance your content.)

As you begin to write lab reports, your instructors will likely ask you to use simple tables and graphs to illustrate your data. This chapter presents some examples of the types of graphs that you will use in your first- and second-year assignments. In particular, the examples focus on ways to graph frequency distributions (relative data displays), some types of longitudinal information (e.g., changes or variations over time), and associations between variables. These examples merely skim the surface of the ways that you can use illustrations to enhance your research, but they provide a good starting point.

When writing a report, always consider how you can present important experimental information graphically rather than verbally. Here are a few basic guidelines for using visual aids effectively:

- Information in an illustration should stand alone. It should be both complete and obvious.
- Information in the text should not simply duplicate information in the illustration.
- Simple illustrations are better than cluttered ones. The easier it is for the reader to grasp the information quickly and accurately, the better.
- As in report headings and subheadings, try to make the title of the visual reflect the point of the illustration, not just the topic. When this is not possible, at least be specific about the content. For example, a title such as "Copepod Density along a Shoreline in Autumn at St. Andrews Biological Station" or "Shoreline Gradient of Copepods" is better than "Copepods in St. Andrews."
- Refer to every illustration in the text, explaining what it shows. If you have several illustrations, number each one so that you can refer to it by number in the discussion.

Frequency Distributions

Before you begin to create tables and graphs, you should understand what a frequency distribution represents. A *frequency* is the number of times a particular occurrence is observed within a specific set of measurements (a data set). A *frequency distribution* is a summary, in the form of a table or a graph, of all frequencies that occur within a data set. As you analyze your data, you should create tables and graphs to help you identify and interpret frequency distributions in different ways.

In your report, you may choose to include a table or a graph (most often a line graph or a histogram) to summarize your frequency data. The choice will depend on the amount and type of information; generally, you should use a table rather than a graph when you are dealing with many categories.

Tables

A table can convey a large amount of information, both numerical and verbal, without losing detail. If you are giving specific information in numerical form, a table allows you to show precise data more clearly than a graph or a chart does. A table is often the best way to display data when small differences in treatments are critical, when there are too many relationships to display in a graph, or when some or all of the information is verbal. Table 10.1 illustrates the type of information that is best represented in a table. In this example, frequencies are summarized by qualitative statements (*common, frequent*, etc.).

Tables are most effective when the columns and the rows are well structured, when white space is used to separate the data, and when the title of the table is clear. It is better to structure a table by frequency than by category labels.[1] Consider the following examples. Table 10.1 lists bat species alphabetically, forcing the reader to reread the frequency column to make sense of the information. Table 10.2 lists the same information, but this time the bat species

Table 10.1 Current extent of bats in Northumberland, England (information organized in alphabetical order by species)

Species	Frequency
Brandt's bat (*Myotis brandtii*)	Rare
Brown long-eared bat (*Plecotus auritus*)	Frequent
Common pipistrelle (*Pipistrellus pipistrellus*)	Common
Daubenton's bat (*Myotis daubentonii*)	Frequent on water
Leisler's bat (*Nyctalus leisleri*)	Rare
Nathusius' pipistrelle (*Pipistrellus nathusii*)	Rare
Natterer's bat (*Myotis nattereri*)	Uncommon
Noctule bat (*Nyctalus noctula*)	Scattered
Soprano pipistrelle (*Pipistrellus pygmaeus*)	Common
Whiskered bat (*Myotis mystacinus*)	Uncommon

Source: Data from Northumberland Biodiversity Partnership, "Northumberland Biodiversity Action Plan: Bats Species Action Plan" (Northumberland Biodiversity Partnership, 2008), https://www.nwt.org.uk/sites/default/files/files/Bats.pdf.

Table 10.2 Current extent of bats in Northumberland, England (information organized in order of decreasing frequency)

Species	Frequency
Common pipistrelle (*Pipistrellus pipistrellus*)	Common
Soprano pipistrelle (*Pipistrellus pygmaeus*)	Common
Brown long-eared bat (*Plecotus auritus*)	Frequent
Daubenton's bat (*Myotis daubentonii*)	Frequent on water
Noctule bat (*Nyctalus noctula*)	Scattered
Natterer's bat (*Myotis nattereri*)	Uncommon
Whiskered bat (*Myotis mystacinus*)	Uncommon
Brandt's bat (*Myotis brandtii*)	Rare
Leisler's bat (*Nyctalus leisleri*)	Rare
Nathusius' pipistrelle (*Pipistrellus nathusii*)	Rare

Source: Data from Northumberland Biodiversity Partnership, "Northumberland Biodiversity Action Plan: Bats Species Action Plan" (Northumberland Biodiversity Partnership, 2008), https://www.nwt.org.uk/sites/default/files/files/Bats.pdf.

are grouped according to frequency, with added white space to separate each group. As you can see, Table 10.2 displays the information more effectively.

Bar Graphs

A bar graph allows you to compare distinct elements at fixed points in time. The data represented in a bar graph can also be displayed in a table, but graphs add a visual element that draws attention to the relationships among the data. As an example, compare Table 10.3 with Figure 10.1—both illustrate the same information, but the bar graph makes a stronger, more immediate impression.

When creating a bar graph, you should always set the baseline at zero to avoid misrepresenting the value of the bars. You can set the bars to be horizontal or vertical, depending on the range of data, and you can make them segmented (stacked) to show different parts of the whole. You can also cluster or group the bars to compare one category with another.

While bar graphs can represent a wide variety of information, they have their limits. They don't work well for displaying highly detailed information, and they cannot clearly show *change* within a data set. If you need to show change over time or trends that run in different directions, you would do better to choose a table or a different type of graph.

Table 10.3 Number of cases of hepatitis C in 2013 in Canada, by province or territory

Province	Number of cases
Ontario	4,156
British Columbia	2,105
Alberta	1,262
Quebec	1,246
Saskatchewan	630
Manitoba	308
Nova Scotia	281
New Brunswick	197
Others	194

Source: Data from Public Health Agency of Canada, "Report on Hepatitis B and C in Canada: 2013," https://www.canada.ca/en/public-health/services/publications/diseases-conditions/report-hepatitis-b-c-canada-2013.html#c3.2

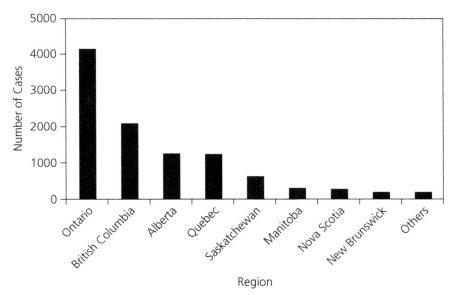

Figure 10.1 Number of cases of hepatitis C in 2013 in Canada, by province or territory

Source: Data from Public Health Agency of Canada, "Report on Hepatitis B and C in Canada: 2013," https://www.canada.ca/en/public-health/services/publications/diseases-conditions/report-hepatitis-b-c-canada-2013.html#c3.2

Histograms

A histogram is useful when you want to display a progression of values across a single category. Whereas the bars in a bar graph can represent distinct categories (e.g., different regions) or non-consecutive measurements (e.g., measurements taken every second year), the bars (or bins) in a histogram represent continuous values of a single variable. The frequency distribution of a histogram may be symmetric or asymmetric (if asymmetric, we call it *skewed*). The bins in a histogram are side-by-side (see Figure 10.2), not separated from one another as the bars are in a bar graph. This positioning gives greater clarity to the data. However, the clarity depends on the number of intervals you choose to use—too many intervals may make the histogram seem cluttered while too few may obscure differences within a bin. There are published rules for working out how to distribute data into bins (e.g., Sturges' rule of thumb), but you should build up your judgment with experience before you consult such guides. Table 10.4 presents the same information as Figure 10.2, but the histogram clearly illustrates the progression of values.

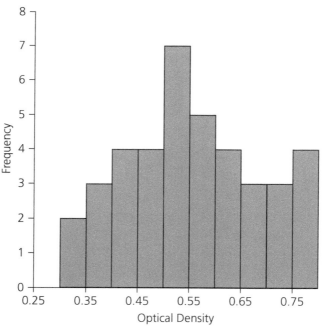

Figure 10.2 Biology 101 class lab results of optical density of water samples from Mud Creek, July 15

Table 10.4 Biology 101 class lab results of optical density of water samples from Mud Creek, July 15

Optical Density	Frequency
0.30	2
0.35	3
0.40	4
0.45	4
0.50	7
0.55	5
0.60	4
0.65	3
0.70	3
0.75	4

Line Graphs

A line graph (see Figure 10.3) shows change over a period of time. It's often used to point out trends or fluctuations. In devising a line graph, put quantities on the vertical axis and time values on the horizontal axis. Try to shape the dimensions of the graph to give the most accurate impression of the extent of change. Never distort your graph to emphasize a point—for instance, by shortening the horizontal axis and lengthening the vertical axis to make a gradual rise look more dramatic. Doing so will only reduce your credibility and cause your reader to question the reliability of your information and the validity of your arguments.

Box Plots

Also known as a box-and-whisker plot, a box plot (see Figure 10.4) is exploratory in nature. It does not assume any statistical distributions or sophistication. It highlights the median value (represented by the line that bisects the box) and displays the remaining data by quartiles (25 per cent of the data). The box represents 50 per cent of the data (the range between the twenty-fifth percentile and the seventy-fifth percentile). You can gauge the departures from symmetry in the data by the relative position of the median. The whiskers represent the remaining data that falls within the distribution (e.g., the data within one standard deviation above or below the mean). Although whiskers can represent other types of distributions, such as ninety-fifth or ninety-eighth percentiles, data found beyond the whiskers are considered outliers in all cases.

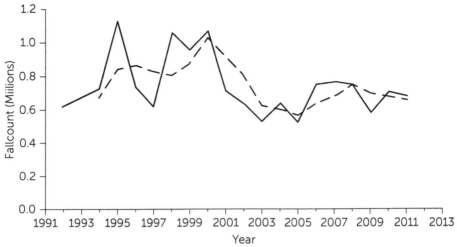

Figure 10.3 Line graph of fall survey numbers of greater white-fronted geese in mid-continent staging areas in Saskatchewan and Alberta

Modified from page 70 of Canadian Wildlife report. http://publications.gc.ca/collections/collection_2012/ec/CW69-16-34-2011-eng.pdf

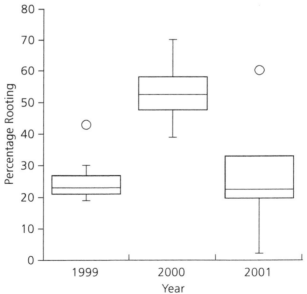

Figure 10.4 Box plot of rooting percentage of juniper cuttings by year

Note: Mild outliers are shown as circles.

Box plots can be very useful before you carry out any analysis because they allow you to see obvious flaws in the data set. A box plot provides an instant view of outlying data points (i.e., mild and extreme outliers, represented by circles in Figure 10.4) that may be due to either variation or measurement errors. Once you identify the errors, you can eliminate them before you perform statistical analysis. Because parametric statistics depends on a number of statistical assumptions that require a data set to be free of errors, you should deal with such errors as early as possible.

Charts such as box plots provide information not only on trends over time but also on data variation. You can assess any given data point within the context of its variation. Graphing trends with variation provides you with ideas on how to test the trends statistically. Even if something looks like a trend, you must determine whether it is statistically significant according to a test. Again, this is an example of how graphing can be useful before analysis. Graphs give you a "feel" for data structure.

Scatter Plots

Sometimes numerical variables can be plotted to establish patterns of association (see Figure 10.5). The resulting cloud of points allows you to think about what kind of curve would best fit the data. You will find an ample number of curve-fitting equations to help you with this task; play with these formulae and always ask whether they capture relevant information.

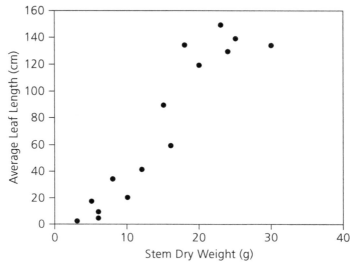

Figure 10.5 Average leaf length versus stem dry weight of ostrich ferns (*Matteuccia struthiopteris*)

Describing Data Presented in a Graph or Figure

Graphs and figures pose problems for students, both when students describe the graph content and when they have to draw a conclusion. Students either unnecessarily repeat data or fail to fully describe the content of the plotted data. In the case of a graph, emphasis must be on interpreting the data, not on repeating the data. Therefore, the graph should be broken down logically. To clarify how you should describe data presented in a graph, Figure 10.5 will be used as an example.

1. The graph axes must be described. For example, "On this graph, the *x*-axis is the stem dry weight in grams, and the *y*-axis is the average leaf length in centimetres."
2. The trends within the data should be noted. This requires some knowledge of statistics. The distribution may be normal or skewed; the trends in the data may show an inversely proportional relationship, or as is the case in Figure 10.5, a proportional rise (e.g., "As stem dry weight increases, leaf length increases"). A graph or figure may also show no discernable trend, that is, the data is randomly distributed.
3. The amount of variability should be noted—for example, the standard error. An objective appraisal of the type and quality of the data is a big help when it comes to interpreting the information. Data can be qualitative or quantitative. In both cases, sample sizes need to be considered.
4. Unusual aspects of the data, such as extreme values, missing values, and missing dates, should also be noted.
5. Once the first four steps have been completed, it is possible to draw a conclusion. Such a statement should summarize what is *shown* in as well as what can be *concluded* from the data. For example, if a rising trend is shown in a graph, but the variation at each data point is so large as to not amount to statistical significance between data points, then it should be concluded that there is no trend. For Figure 10.5, the conclusion is relatively simple: "As ostrich ferns increase in mass, their leaf length, on average, increases proportionately in a statistically significant manner. This means that leaf length is not, as might have been expected, fixed in these ferns, but is correlated with the accumulated resources of the stem."

Pie Charts

A pie chart is used to emphasize proportions—to draw attention to the relative size of the parts that make up a whole. You should avoid pie charts whenever

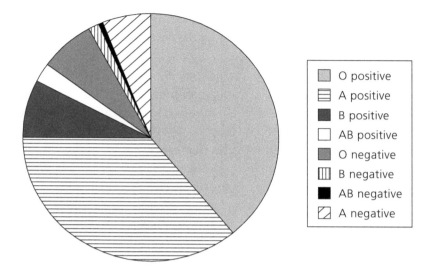

Figure 10.6 Pie chart of blood types in the Canadian population
Data source: Canadian Blood Services.

possible because they have very low data density—the only data structure is the relationship between shape and value. They also have less immediacy than the graphs discussed in this chapter, as they require the reader to look back and forth to the legend to make any sense of the diagram. How do you decide when to use a pie chart? Create a table with the same data. If a pie chart makes a greater immediate impact than a well-ordered table, then a pie chart is called for. However, well-constructed tables are generally superior to pie charts.

Photographs

Photographs can make an assignment more interesting. Selecting a good photograph requires good judgment. The most effective photographs reinforce the key message of the assignment. An illustration can be superfluous if it does not show what is important. For example, if the assignment is about bear fishing behaviour, a picture of the species of bear is far less important than a picture of one bear fishing. If a perfectly relevant picture is unavailable, then it is best not to include a picture; otherwise, the picture itself becomes a distraction. Distracting pictures negatively affect the narrative unity of the assignment, whereas properly integrated visuals add value to an assignment.

The photographs you use should look crisp and clear on the page when printed. Pictures that are well balanced and have excellent contrast are preferred over photographs that are out of focus or have large pixels. Avoid odd-shaped photographs (e.g., round pictures). Also avoid pictures that have oddities in alignment (e.g., off-centre subjects) or in colour composition (e.g., very dark or very light photographs).

A common rule in photography is that the picture should be divisible into thirds—imagine dividing a photograph into three equal parts, with the foreground in one third, the subject matter in one third, and the background or sky in one third. For a general idea of what makes an excellent photograph, turn to magazines and websites that specialize in biological photography. Both *Canadian Geographic* and *National Geographic* have extremely high-quality photographic essays, many of which are on biological subjects.

You can import JPEG, TIFF, and other common image formats into a Word document. In Microsoft Word, find the "Insert" tab on the top menu bar and then select "Picture." Select your picture from the appropriate folder and then click "Insert."

The two most common problems that students encounter in putting pictures in their assignments are resizing and pixelation.

To resize an image in Microsoft Word, follow these steps:

- Left-click once on the image. A frame appears around the image with small circles at the corners and squares at the midpoints (see Figure 10.7).
- Three types of control handles are available. The top handle is for rotation, but the other two controls are relevant for resizing. The small circles and squares are known as sizing handles.
- The circles in the corners of the frame control proportional resizing. Click on one of the circles and drag the handle to change the size of the image smaller or larger. Do not use the squares as these will change the width or height independent of one another, resulting in distortion.
- When the desired size is reached, release the cursor from the handle.

Pixelation problems arise when photographs of low resolution are resized and the pixels become apparent. Figure 10.8 shows an example of how resizing a low-resolution image results in pixelation. The picture on the left looks clear at a small image size. When this image is expanded (see Figure 10.8 on the right), the pixels become obvious and the X-ray micrograph appears to be out of focus. This common error is easily avoided by using pictures that are of high initial resolution. If a picture is not of sufficient resolution to be made larger, there is, unfortunately, no way to improve it. The only choice is to retake or recreate the image at a higher resolution.

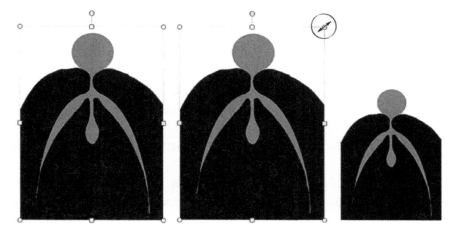

Figure 10.7 Alteration of image size by using the sizing handles

Note: The schematic illustration of a cross section of a *Ginkgo biloba* ovule is reduced by left-clicking on the image and dragging the top right-hand sizing handle.

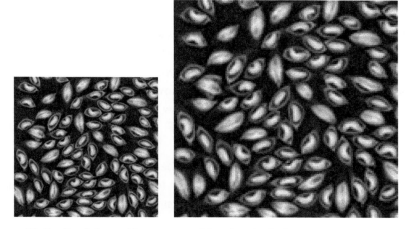

Figure 10.8 Resizing of image resulting in pixelation

Note: The X-ray of Douglas-fir seed parasitized with wasp larvae is clear as sized on the left. But when the image size is increased, the pixels become evident—making the image quality unacceptable (right).

Labelling Graphs and Figures

A figure must be labelled with a caption, which is placed below the figure. In contrast, the title of a table is always placed above the table. A general rule is that the text in a title or caption must stand alone, which is to say, if someone were to open your assignment and read the accompanying text to a table or figure, that information ought to be self-explanatory.

The detail to be included in a caption depends on the nature of the illustration. For example, a photograph usually requires only a simple caption:

> Figure 1. Grizzly bear catching salmon on the Fraser River.

It isn't necessary to write, "Photograph of a grizzly bear catching salmon on the Fraser River."

More technical figures require more details. The description of a highly processed image must also include some additional standard details. The type of illustration should be included (e.g., schematic, model, flow chart, electron micrograph, confocal image, X-ray, etc.). A scale bar or the magnification must also be provided.

> Figure 2. Electron micrograph of a soybean nucleus. Magnification 850 x.
>
> Figure 3. X-ray image of bones in a human hand. Scale bar = 5 cm.

Illustrations that are not your own work must be referenced. For example, the source of illustrations that have been downloaded from online sources, photocopied, or scanned, must *always* be acknowledged in the image caption. The first example below shows how to cite an image used from a book; the second example shows how to cite an image used from an online source:

> Figure 4. Electron micrograph of a soybean nucleus. Magnification 850 x. From: Newcomb W. 2017. Guide to soybean cell ultrastructure. Kingston (ON): McGill–Queen's University Press. p. 22.
>
> Figure 5. Sunflower. Wikipedia. 2009 Aug 21 [cited 2017 June 2]. Available from: http://en.wikipedia.org/wiki/File:Sunflower_sky_backdrop.jpg.

If a figure comes from a source, the source can simply be listed in the figure caption. However, if the same source supports an argument in the body text, it must then also be acknowledged in the bibliography or literature cited section.

Graphing Software

The number of ways to display data has increased in the last 20 years, thanks to developments in computing. Computer programs make it easy to create, format, and annotate tables, graphs, and diagrams that illustrate important points in your essay or report. Often, you can create a variety of illustrations based on a single data set. Spreadsheet programs such as Excel allow you to create simple charts and graphs, but you will likely have to use more specialized programs to analyze more complex data. If you want to create a box plot, for example, you will need to use statistical software such as Minitab (a fairly simple suite of statistical programs) or SPSS (a more complex statistical program). You should be able to access programs such as these through your school's computing or statistics department. You can also use R, an excellent graphing software package that is free to download. In addition, some universities require their students to master MATLAB.

Tables or Graphs?

When deciding whether to use a table or a graph to present your data, you must always consider which will be the most effective in getting your point across. Tables have the advantage of being more exact than graphs because they provide precise numerical information, and they can include text as well as numbers. Tables are also a good choice if you have several sets of numbers that could get buried if you just listed them in the text. However, if you have data for a number of conditions that vary systematically, a graph is the best way to illustrate the information. Graphs are useful for drawing attention to relationships among the data. When the structure of the data is simple, a bar graph may do the job, but when the structure is more complex, you should use a more complex type of graph.

You might also want to consider practical implications, such as which method of presentation uses the most pixels, takes up the most space, or uses the most ink when printed. Tables are usually the most economical option, in terms of both space and printing cost, followed by box plots and line graphs; bar graphs, pie charts, and histograms can take up a lot of space and consume a lot of ink. In the end, the format you choose will often come down to nothing more than personal style or preference.

Of course, you don't always need to present your data in a graph or a table. If you're making a simple comparison in a lab report, for example, you can include your data analysis directly in the text of the *Results* section:

> Understory trees in the 10–15 year age class were shorter in shaded conditions (11 ± 2 m) than those in sunny conditions (18 ± 3 m, p < 0.001).

Three Dangers in Using Illustrations

- Computer programs offer a wide range of features and options for tables and graphs. Be sure that the designs you create are not too elaborate for your purpose. Visuals are meant not to dazzle but to make it easier for the reader to understand. In general, keep your style simple and avoid stacked bar graphs, 3-D bar graphs, and pie charts. Clarity and immediacy must always be the goal.
- Any illustration, even if it is created on a computer, can distort information. For instance, the slope of a graph line can be made to look steep or shallow depending on the graph's scale. Trend lines can be made to begin at a time point that omits unfavourable periods. Although line and bar graphs are the most susceptible to distortion, the shapes and proportions of other diagrams can also give a false picture. Figure 10.9 shows a typical example of such a distortion—both graphs present the same data, but the increase in population is made to appear much greater in the graph on the right, because the vertical axis does not begin at zero. Be careful to present as accurate a picture as possible so that your illustrations reinforce

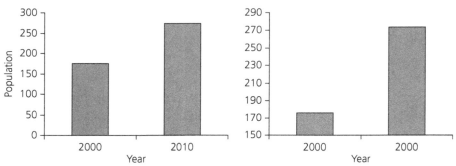

Figure 10.9 Example of misleading graphics

Note: The increase in the population of whooping cranes is made to appear much more dramatic in the graph on the right compared with that on the left. Both are accurate, but the y-axis of the graph on the right begins at 150, which gives the impression that the increase is greater than it really is.

the credibility of your words. A good test of a graph or table is whether you can imagine how the data that went into the illustration was collected. A good illustration is one that allows you to understand the decision-making process that led to the creation of the graph, table, or chart.
- You might think that you can use a picture of enzyme-folding or a graph from a scientific journal for free. After all, projects for elementary school and high school allow images to be cut from magazines or websites and pasted. University or college is the time to leave behind these practices. Lab reports and essays are not projects but assignments created in a scientific tradition. Consequently, you must acknowledge the source for every figure and for all data that is not your own in order to avoid the charge of plagiarism. If the material is restricted by copyright, you will need to contact the copyright holder and ask for permission. If you have any questions, consult your school's policies on copyright or the federal government's Copyright Act. Reference librarians should also be able to answer any questions that you may have about using materials under copyright. Often, the best way to include information from an external source in your assignment is to recreate it in your own style, reference the source, and include the statement "modified from . . ." in a note below the illustration—for example, "modified from Grant et al. 2017."

Summary

Converting raw data into a table or a graph is an essential step in data interpretation. Illustrations such as these allow you to get a feel for the data before you develop your analysis. When writing a lab report or an essay, let your textual descriptions flow from the illustrations and not vice versa. To be successful, illustrations must be easy to understand, so you should avoid using unnecessarily elaborate formatting. The tables, figures, and simple graphs covered in this chapter will help you fulfill the requirements of first- and second-year courses, but you should learn to use more complex models such as flow charts and diagrams once you are familiar with the basics. As your studies progress, you will need to analyze more complex data through more complex illustrations. To prepare yourself, begin to learn different types of graphing software that can handle more and varied types of data.

Note

1. Gerald van Belle, *Statistical Rules of Thumb* (New York, NY: John Wiley and Sons, 2002), 221.

11 Documenting Sources

Objectives
- Understanding the purpose of documentation
- Using CSE style in a life sciences essay
- Using CMS style in a life sciences essay
- Developing an annotated bibliography
- Avoiding secondary sources
- Using reference managers

Much of the writing you do will require you to consult sources to become familiar with current research and to find support for your ideas. It is essential to acknowledge those sources, not only when you quote directly from them but also when you restate, in your own words, arguments or ideas taken from them. If you do not acknowledge your sources, you are allowing your reader to assume that the words, ideas, or thoughts are yours; in other words, you are plagiarizing, and the penalties can be severe.

The purpose of documentation is not only to avoid charges of plagiarism but also to show the body of knowledge that your work is building on. Academic writing is based on the premise that researchers are not working in a vacuum but are indebted to those scholars who came before them. By documenting your sources, you are showing that you understand this concept and are ready to make your own contribution to the body of knowledge in your field.

Understanding Why Documenting Sources Provides Academic Integrity

Life sciences students get their information from many sources. At one extreme, information is passed along that may not be reliable; this happens in conversations, newsfeeds, and social media. Slightly more reliable information is found in newspapers and popular science magazines. Most reliable are academic reviews and peer-reviewed literature, such as articles in scientific trade journals. At each and every level mentioned above, sharing information is easy. Sharing information in science is equally easy, if well-known conventions concerning sources are adhered to. A student's academic integrity is considered to be high if they master these conventions of documenting sources. Equally, a student's academic integrity will be significantly damaged if they do not conform to these conventions.

Plagiarism, in the case of documentation, means passing off someone else's work as your own by omitting the source reference. Copying ideas, sentences, paragraphs and more without attribution is the grossest form of plagiarism. Omission need not be so deliberate: omitting a source often occurs because of a failure to pay attention to detail (i.e., sloppiness). Unfortunately, students are not usually considered innocent until proven guilty. Whether it is a professor or a teaching assistant who is tasked with marking carelessly produced written work, they will have trouble distinguishing sloppy omissions from omissions by far more dishonest students (i.e., students who set out to mislead). Because the charge of plagiarism is probably levelled at more students than is warranted, you would be wise to remember that the *onus is on* you *to document your sources meticulously*.

Understanding Why We Need Documentation Styles

If you were to write an assignment that contained only one quotation, one reference, or one figure based on an external source, it would hardly seem necessary for you to follow an elaborate system of rules when it comes to referencing. But most essays and lab reports involve a dozen or more references. Without stylistic consistency, a writing assignment with many references would likely frustrate the reader as he or she tried to sort through haphazardly presented references to get to the purpose of the assignment.

The purpose of an essay is to explore a theme or a thesis; the purpose of a lab report is to explore a hypothesis. External sources of information, or references, are not central to the purpose of these writing assignments—they serve only to bolster the arguments. Consequently, writers try to draw attention away from references by reducing their visibility on the page. A citation style is a system that allows writers to keep their writing clear by using an in-text shorthand to refer readers to a reference, usually listed in full at the end of the document.

A consistent style of citation also forces the writer to include all information that the average reader would need in order to retrieve the cited source.

Many professionals, academics, and students like to record a selection of references as they read an essay because these articles may prove to be interesting. In professional articles, references can not only provide evidence for an argument but also, on occasion, point the reader to an article or study that has been forgotten, neglected, or underappreciated.

Should formatting be a priority for an undergraduate? Yes! Because assignments at all levels in science require precise formatting of citations and references, it is very easy to separate the students who pay attention to detail from those who do not. Before marking an essay, a professor or TA need only look at the quality of the formatting in the reference list to get a pretty accurate idea of whether the student is careful and meticulous, or rushed and sloppy. Most of the time, the quality of the bibliography can be used to predict the quality of the essay.

Documenting Your Sources

There are many systems of documentation. The style you use will depend on the subject you are writing about as well as the preference of your instructor, your department, or your employer. Students who write essays for English follow the guidelines set out in the *MLA Handbook for Writers of Research Papers*. Students who write essays for psychology adhere to the rules of the *Publication Manual of the American Psychological Association*. In the life sciences, writers follow the guidelines set out in *Scientific Style and Format*, published by the Council of Science Editors, and *The Chicago Manual of Style*, published by the University of Chicago Press. (Both CSE style and Chicago style are discussed in detail in this chapter.)

In addition, some science journals use their own style of referencing. If you are asked to write your essay following the style of a particular journal, then you will have to look at articles published in that journal to determine that journal's conventions. A journal's style requirements are found on a page entitled "Instructions to Authors" or something similar. You will find that most journal-specific styles are similar to those discussed in this chapter. If you are not required to follow a specific documentation style, you should consider using either of the two most common systems of documentation used in the life sciences.

Note also that the latest version of your word-processing software is likely able to automatically format endnotes, footnotes, and citations as well as bibliographies and lists of works cited according to CSE and Chicago styles. It's worth your while becoming familiar with these features, as they are fast and accurate and can save you a great deal of time and effort.

CSE Style

The Council of Science Editors (CSE) is the recognized authority on documentation styles in all areas of science and related fields. The following guidelines are based on *Scientific Style and Format: The CSE Manual for Authors, Editors, and Publishers* (8th ed., 2014).

The CSE manual outlines three major systems for documenting sources: name–year, citation–sequence, and citation–name. A brief description of each system follows.

Name–year

In the name–year system, in-text references consist of the surname of the author and the year of publication enclosed in parentheses. Complete bibliographical information is given in a list of references, organized alphabetically by author surname, at the end of the paper. In the following examples, note that CSE style emphasizes simplicity, avoiding formatting such as italics and minimizing the use of punctuation.

In-text reference

> Differences in ovule structure have been used to separate modern seed plants into either angiosperms or gymnosperms (Tomlinson 2012).

End reference

> Tomlinson PB. 2012. Rescuing Robert Brown—the origins of angio-ovuly in seed cones of conifers. Bot Rev. 78(4):310–334.

Citation–sequence

In the citation–sequence system, superscript numbers in the text correspond to numbered references in a *References* or *Cited References* section at the end of the document. These references are listed in the order in which they are first cited in the text. Thus, the first reference used in the text will be 1, the next new reference cited will be 2, and so on. Once a source has been assigned a number, it is referred to by that number wherever it appears in the text:

> Common contaminants found in drinking water in Indigenous communities in Ontario include uranium[1], trihalomethanes[2] and *Escherichia coli* [3]. Trihalomethanes are by-products of disinfectants used to control microbial contaminants[1,4].

If several sources are being referenced at once, all relevant reference numbers should be given in the same citation. Reference numbers should be separated by commas with no spaces. When using a sequence of three or more citation numbers (e.g., 7, 8, 9), use only the first and last number in the sequence, separated by a hyphen:

> Several studies[1,5,12–15] have shown . . .

In-text reference

> Self-incompatibility (SI) prevents inbreeding through specific recognition and rejection of incompatible pollen[1].

End reference

> 1. Behdarvandi B, Guinel FC, Costea M. Differential effects of ephemeral colonization by arbuscular mycorrhizal fungi in two *Cuscuta* species with different ecology. Mycorrhiza. 2015;25(7):573–585.

Citation–name

In the citation–name system, the list of end references is compiled alphabetically by author surname. The references are then numbered in that sequence, with Aaronsen number 1, Babcock number 2, and so on. These numbers are used for in-text references regardless of the sequence in which they appear in the text. If Mortensen is number 38 in the reference list, the in-text reference is number 38, and the same number is used for subsequent in-text references.

When several in-text references occur at the same point, place their corresponding reference-list numbers in numerical order. In-text reference numbers not in a continuous sequence are separated by commas with no spaces. For more than two numbers in a continuous sequence, connect the first and last with a hyphen.

> . . . are illustrated in studies[2,7–11,16,25] that corroborate . . .

Formats for end references are similar to those used in the citation–sequence style as shown above.

End references

The following examples illustrate end references for the citation–sequence and citation–name systems of documentation. The name–year system differs by placing the year of publication directly after the author name(s).

Note that if you cite material from a publication that you have not read but have seen cited by others, you should cite the source that you actually read; do not cite the original (see pages 201–202 for more details).

What do the parts of the end reference mean?

Following the note number (used for the citation–sequence and citation–name systems), an end reference starts with the names of the authors—beginning with the lead author's last name, followed by his or her initial(s) and the names and initials of additional authors. In the following example, the journal article has four authors:

> 1. Champredon D, Cameron CE, Smieja M, Dushoff J. Epidemiological impact of a syphilis vaccine: a simulation study. Epidemiol Infect. 2016; 144(15):3244–3252.

The authors agreed to have their names on this paper because each one contributed ideas, labour, or essential resources that resulted in this successful study.

The title of the article is meant to be informative: in this example, the article is about a mathematical model used to explore the potential impact of rolling out a hypothetical syphilis vaccine. Because the nature of this work is epidemiology, the paper was published in a well-established infectious disease journal, *Epidemiology and Infection*, which is abbreviated in a standardized manner to Epidemiol Infect. The date, 2016, is followed by the volume number, 144, and the issue number, (15), of the journal. The reference ends with the page range or extent; this article is 9 pages long (pages 3244–3252).

Notes on references—essential and nonessential types

Publications often add information to improve a reader's ability to download an article in a reference list. These notes are added to the end and may provide either nonessential or essential information. Nonessential notes are not required to find a reference; that is, the reference minus the note would still be perfect. Such notes may include, for example, references to the language of the article, its availability, and in the case of technical references to software, platforms suitable for running the software.

In the example below, we have added a non-essential note to Champredon et al. 2016. It now has a DOI (Digital Object Identifier). The reference without the DOI was perfectly adequate, but with the addition of the DOI, a researcher can call up the paper more quickly, perhaps, than without it. Clicking on the DOI directly if it was an active link, as is often the case in online papers, would direct the reader to the journal article. Alternately, the DOI could be copied and pasted into a browser that would direct the reader to the source article. DOIs are assigned by an international registry foundation to create persistent links to any object's location on the internet. The DOI always begins with a 10 followed by a prefix of four digits that is assigned *to the organization* (e.g., publisher), and a suffix that is assigned *by the organization* that is much longer and is designed to be flexible to meet, for example, the publisher's internal identification requirements.

1. Champredon D, Cameron CE, Smieja M, Dushoff J. Epidemiological impact of a syphilis vaccine: a simulation study. Epidemiol Infect. 2016;144(15):3244–3252. doi:10.1017/S0950268816001643

A common example of essential information added to the end of a reference is a Uniform Resource Locator (URL). As publishing becomes ever-increasingly paperless, more and more reports from governments and non-governmental organizations are posted online. Without the URL, they would be impossible to access. This is why these notes are considered to be essential notes. Unlike a DOI—a permanent record that can always be retrieved—a URL is a link to an online resource that is not permanent. Take the following example. A government report may be relocated to another server when ministries are merged or altered in their structures. This means that the report is moved from one server to another, thereby changing its locator or address (i.e., its URL). A URL may be changed as resources are moved even within a server, discarded, or archived.

1. Woods J, Mahony C. Climate-based seed transfer: Guiding British Columbia's reforestation investments in a changing climate. Forest Genetics Council of British Columbia. Victoria BC. 2016. 5 p. http://www.fgcouncil.bc.ca/CBST-investment-extension-note-FINAL.pdf

Book with one author

1. Gregg J. Are dolphins really smart? The mammal behind the myth. Oxford (UK): Oxford University Press; 2013. 320 p.

Note that the last element of the entry ("320 p.") indicates the total number of pages in the book. Although this is an optional component of a book reference, it can provide useful information to the reader

Book with two or more authors
If the book you are referencing has more than one author, give the names of all authors up to a maximum of 10; after 10, replace additional authors' names with "et al." Names should be inverted and separated by commas. Note that "and" is not used:

2. Withers PC, Cooper CE, Maloney SK, Bozinovic F, Cruz-Neto AP. Ecological and environmental physiology of mammals. Oxford (UK): Oxford University Press. 2016. 560 p.

Book by an organization as author

3. National Advisory Committee on Immunization (CA). Canadian immunization guide. 7th ed. Ottawa (ON): Public Health Agency of Canada; 2006. 379 p.

Book with an editor in place of an author

4. Berta A, editor. Whales, dolphins and porpoises. Chicago (IL): University of Chicago Press; 2015. 288 p.

Chapter or other selection in a book

5. Viney ME. Life-history plasticity and responses to host defence. In: Kennedy MW, Harnett, W. editors. Parasitic nematodes: molecular biology, biochemistry and immunology. Wallingford (UK): CAB eBooks. 2013. p. 30–41. doi:10.1079/9781845937591.0000

Article in a journal

   6. Bateman AW, Anholt BR. Maintenance of polygenic sex determination in a fluctuating environment: an individual-based model. J Evol Biol. 2017;30(5):915–925.

Note that CSE abbreviates the names of journals.

Article in a magazine

   7. Popescu A. How do LA's pumas cross the road? New Scientist. 2017 1 July: 10.

Article in a newspaper

   8. MacDonald G. CBC on trial. Globe and Mail. 2001 Jul 28;Sect. R:12 (col. 4).

   9. Rare skunk injured in trap is put down. Times Colonist. 2013 Nov 3;Sect. A:7 (col. 3).

Lecture or paper presented at a meeting

   10. Dickinson TA. Carrie Derrick—botanist and Canada's first woman professor. Canadian Botanical Association 53rd Annual Meeting; 2017 Jul 4–8; Waterloo, ON.

Article in an online journal

   11. Shen X, Husson M, Lipshutz W. Chronkhite. Canada syndrome: a case report and literature review of gastrointestinal polyposis syndrome. Case Rep Clin Med. 2014 [accessed 2017 July 5];3(12): 650–659. http://ww.scirp.org/journal/Paperinformation.aspx?PaperID=52581. doi:10.4236/crcm.2014.312138

The article above has both the URL and the DOI.

Website

12. Dementia in Canada. Ottawa (ON): Public Health Agency of Canada; 2017 [updated 2017 Jun 28; accessed 2017 Jul 29]. www.canada.ca/en/public-health/services/publications/diseases-conditions/dementia.html

Chicago Style

The Chicago Manual of Style (17th ed., 2017) outlines two methods of documentation:

1. The notes and bibliography method, also known as the humanities style, is preferred by those in literature, history, and the arts. It uses superscript numerals to direct the reader to footnotes at the bottom of the page or endnotes on a separate page at the end of the document.
2. The author–date system, preferred in the physical, natural, and social sciences.

For more information on either method, consult the manual itself or the online "Chicago-Style Citation Quick Guide" (http://www.chicagomanualofstyle.org/tools_citationguide.html), which gives examples of CMS references.

Author–date system

The following sections outline the author–date system, broken down into in-text citations (T) and reference-list entries (R). A page number or other locator can be added after a comma. Although this is uncommon in life sciences, it may help direct a reader to a critical piece of information.

Book with one author

T: (Eigen 2013)
R: Eigen, Manfred. 2013. *From Strange Simplicity to Complex Familiarity: A Treatise on Matter, Information, Life, and Thought*. Oxford: Oxford University Press.

Book with two or three authors

    T: (Torday and Rehan 2017)
    R: Torday, John, and Virender Rehan. 2017. *Evolution, the Logic of Biology*. New York: Wiley.

Note that with four or more authors, in-text references give just the first author followed by "et al." The reference list, however, lists all authors' names.

Book with an organization as author

    T: (FAO 2015)
    R: FAO (Food and Agriculture Organization). 2015. *70 Years of FAO (1945–2015)*. Rome: Food and Agriculture Organization of the United Nations.

Book with editors in place of authors

    T: (Cacioppo et al. 1987)
    R: Cacioppo John T., Louis G. Tassinary, and Gary G. Berntson, eds. 2017. *Handbook of Psychophysiology*. Cambridge: Cambridge University Press.

Book with an editor or translator in addition to an author

    T: (von Uexküll 2010)
    R: von Uexküll, Jakob. 2010. *A Foray into the Worlds of Animals and Humans: With a Theory of Meaning*. Translated by Joseph D. O'Neil. Minneapolis: University of Minnesota Press.

Chapter or other selection in a book

    T: (Ferry 2011)
    R: Ferry, Michel. 2011. "Potential of Date Palm Micropropagation for Improving Small Farming Systems." In *Date Palm Biotechnology*, edited by Shri Mohan Jain, Jameel M. Al-Khayri, and Dennis V. Johnson, 15–28. New York: Springer.

Article in a journal

T: (Williams 2012)
R: Williams, Joseph H. 2012. "Pollen Tube Growth Rates and the Diversification of Flowering Plant Reproductive Cycles." *International Journal of Plant Sciences* 173, no. 6: 649–61.

Article in a magazine or newspaper

T: (Dunietz 2017)
R: Dunietz, Jesse. 2017. "Nevermore, or Tomorrow? Ravens Can Plan Ahead." *Scientific American*, July, 2017.

Lecture or paper presented at a meeting

T: (Berland 2014)
R: Berland, Anne. 2014. "Climate Impacts on Seed and Cone Production in Provenance Trials in British Columbia." Paper presented at the Annual Conference of the British Columbia Seed Orchard Association, Salmon Arm, British Columbia, Canada, June 17–18, 2014.

Article in an online journal

T: (Chan et al. 2012)
R: Chan, William F.N., Cécile Gurnot, Thomas J. Monine, Joshua A. Sonnen, Katherine A. Guthrie, and J. Lee Nelson. 2012. "Male Microchimerism in the Human Female Brain." *PLoS ONE* 7, no. 9 (2012): e45592. https://doi:10.1371/journal.pone.0045592.

When citing a journal article that was accessed online, include the digital object identifier (DOI) at the end of the reference list entry. If a DOI is not available, include the URL at the end of the entry.

Website

Websites may be cited in running text instead of in an in-text citation ("The CBC website lists several . . ."). Websites are often omitted from reference lists as well. If you are required to give a more formal citation, follow the examples below. If a site was updated/modified, provide that date. Otherwise, provide the date that you accessed the site. Frequently, no date of publication or revision is available; in such cases, use n.d. (no date), in small letters.

- T: (CBC 2014)
- R: CBC. 2014. "DFO Extends Cod Fishing Season 'Indefinitely' on South Coast." *CBC News*. Last modified March 11, 2014. http://www.cbc.ca/news/canada/newfoundland-labrador/dfo-extends-cod-fishing-season-indefinitely-on-south-coast-1.2568149.
- T: (National Plant Monitoring Scheme, n.d.)
- R: National Plant Monitoring Scheme, n.d. Botanical Society of Britain and Ireland (website). Accessed November 10, 2017. http://bsbi.org/npms.

Blog

As in the case of websites, blogs are often cited in running text only ("In a comment posted on the *New Bedford Whaling Museum Blog* on November 1, 2016 . . .") and are omitted from a reference list. If you are required to provide more formal documentation, follow the example below. If an access date is required, include it before the URL.

- T: (Dyer 2016)
- R: Dyer, Michael P. 2016. "Why Black Whales are called "Right Whales". *New Bedford Whaling Museum Blog*, September 2016. http://whalingmuseumblog.org/.

Writing an Annotated Bibliography

Some research assignments require an annotated bibliography, which is a standard list of sources accompanied by descriptive or evaluative comments on each item in your list.

If you are asked for an annotated bibliography, begin by arranging your list of entries just as you would for a standard bibliography. Then include a brief comment about the source—for example, the quality of the information it contains, the approach of the author, a brief analysis of the paper's strengths and/or weaknesses, or its relevance to your subject.

Consult whatever style guide you are using for details about the specific format recommended for an annotated bibliography.

Avoiding Indirect References

A common mistake in essays and lab reports is referring directly to a work that has been quoted in one of your reference sources but that you haven't actually read. Such a reference is indirect, or secondary. Whenever possible, you should find the original source, read it, and make your own assessment of the material. Occasionally, you will find that a critically important reference is unavailable or it is in a foreign language that you cannot read. In either case, you must reference both the original source and the source in which you read the quotation.

For example, if Östergren's work is cited in a paper by Ågren but you did not read Östergren's original paper, you must credit your source—Ågren:

> Östergren (as cited in Ågren 2013) was one of the first to think of selfish genes . . .

In the reference list, include an entry for your source (Ågren).

In the case of a frequently cited classic work written in another language, you need to be careful to reference the correct source. If you are writing about the evolution of plant form, you may wish to refer to Goethe's ideas on adaptability of form. In a modern book on the history of botany by A.G. Morton, you found a quotation that you would like to use in your assignment. You noticed that Morton has quoted Goethe's 1790 book *Versuch die Metamorphose der Pflanzen zu erklären*. If you were using Chicago's author–date system for your citations, you would mention Goethe in your discussion and use the following format for citation:

> Goethe identified principles of organization in above-ground lateral plant parts: "Whether the plant produces leaves, flowers or fruit, it is always only the same organs which, in manifold determinations and often in changed forms, fulfill nature's demands" (quoted in Morton 1981, 474).
>
> Reference: Morton, A.G. 1981. *History of Botanical Science*. New York: Academic Press.

You should never include in your reference list an entry for a document that you have not actually seen. If you are using one of the styles outlined by the CSE and you cannot avoid referencing such a document, cite the source in which you found the quotation:

As Morton has noted, Goethe had identified principles of organization in above-ground lateral plant parts as early as 1790: "Whether the plant produces leaves, flowers or fruit, it is always only the same organs which, in manifold determinations and often in changed forms, fulfill nature's demands."[1]

Reference: 1. Morton AG. History of botanical science. New York (NY): Academic Press; 1981. 474 p.

Reference Managers

Reference managers are software used to publish and to manage bibliographic data. They shorten the time needed to create literature citation lists in essays and lab reports. They also allow life sciences students to build and to manage a collection of scientific references on any subject that may interest them.

Space does not permit us to provide a step-by-step guide; however, reference managers are relatively easy to master. They will spare you a lot of work over the course of a degree. Every Canadian university and college library that we searched not only provides access to reference managers but also provides online lessons on how to get started. The most common are Zotero, EndNote, and Mendeley, but others are also available. A number are also available either by subscription or as freeware.

Summary

Documentation can take various forms depending on the discipline, but all documentation styles share some common benefits. A consistent documentation style makes it easy for a reader to retrieve a source, allowing him or her to verify the facts and opinions you have obtained from external sources. A consistent style also gives your assignment an organized, professional look, making the content more readable. In the life sciences, the most common styles are Chicago and CSE. These styles cover everything from formatting quotations to adding citations and references to organizing footnotes and endnotes. You may feel overwhelmed by style points the first time you confront such a battery of conventions, but with practice, you will save time. Ultimately, using a style manual will make your life easier because you won't have to spend time deciding on the best way to cite a particular piece of information—the guide makes all of the decisions for you.

12 Giving Oral Presentations and Poster Presentations

Objectives
- Preparing and delivering a talk
- Answering questions
- Handling pressure
- Improving your delivery—using a script
- Creating effective visuals
- Preparing a poster
- Improving your poster graphic design
- Learning by example—three sample posters

To master the art of giving presentations, you must keep in mind the type of audience you will be addressing. When you give a talk to an audience in a room, it is a single presentation to many people, whereas when you are preparing a talk to accompany a poster, you will be giving multiple presentations, usually one-on-one. In both cases, your talk will need to be clear both in its graphics and in its exposition if you are to keep your audience's attention. In the case of the oral portion of your poster presentation, it will essentially take the form of a question-and-response discussion, and you will be able to tailor each interaction to the interests of the individual listener. Therefore, you don't need to structure the oral part of your poster presentation as precisely as you do for a talk. A successful poster is one that is self-explanatory and can be read on its own. However, a successful poster session, in which the poster's creator is explaining its contents, depends as much on having nice graphics as on the presenter's ability to explain his or her scientific experiment.

Oral Presentations

If you think about the similarities that bad seminars and lectures share, you'll probably find that the speaker either jumped from topic to topic, or too quickly passed over crucial elements of his/her argument, or, worse, took for granted information that had never been explained. By comparison, good seminars are more likely to be well organized and systematic, with the speaker leading you through the material being discussed in a logical manner. Some individuals are naturally more comfortable in front of an audience than others. However, you will find that even if public speaking is not one of your natural talents, you can still give a successful presentation by following a few simple rules. The two most important of these are *be prepared* and *be organized*.

Before you plan your presentation, consider the overall process. You can break this into three parts: making preparations, delivering the talk, and responding to questions. The sections below offer advice to guide you through the entire process.

Making preparations

Know your topic
For the purposes of an in-class seminar, you are the expert and will probably know more about your topic than any of the other students. You need to show your audience that your grasp of the subject matter goes beyond what you include in your talk. If you know more than what you present, you will be better able to answer questions. The more background reading you do, the more information you will have to fall back on when someone asks a question. This extra preparation will also boost your confidence.

Consider your audience
Never prepare a seminar based solely on what *you* know about the topic. In fact, you should approach the talk from precisely the opposite direction: put yourself in the position of a listener. If you were sitting in class instead of standing up at the front, what would you expect of the speaker? How much will the typical audience member know about this topic? What will the typical audience member find most interesting? Most relevant? What can you take for granted as common knowledge in the context of the course? If you combine these with your own question—What do I want my audience to know?— then you have the basis for setting up your talk.

Plan your presentation

Giving a presentation involves much more than writing an essay and then reading it out to the class. By the time you are asked to give an in-class presentation, it is likely that you will have sat through dozens of lectures. The best lecturers were the ones who were the most prepared, who spoke without reading directly from their notes, who used a range of visual aids, and who seemed animated and interested in what they were talking about. You can be just as interesting by following some of these suggestions:

- **Decide how much material you need to script in advance**. Some students, particularly those who are already skilled orators, find it useful to fully script their presentations in advance (see pages 213–217). Other students, especially those with less presentation experience, find that writing out their whole talk makes their presentations sound laboured and monotonous. If you are one of these students, draw up an outline that will serve as a guide as you move through your talk. You should also create point-form notes for each of the points that you are planning to discuss. Because you can't read rough notes to your audience, you will be forced to use your own words and, likely, a more natural speaking style. If you are worried that you may freeze when you begin, write out the first paragraph of what you want to say, just to get you started.
- **Prepare an outline for your audience**. An outline will give your audience members something to follow as you talk. Typically, you will base this outline on one that you used to construct your talk, but for the presentation you may also wish to include additional details, such as a list of references.
- **Use visual aids**. Having visual aids serves several purposes. First, visual aids can attract and focus the audience's attention. If you are likely to become self-conscious when standing in front of a group, you will be more at ease when all eyes are on your visual aids and not on you. Second, visual aids provide reminders of what you need to say. If you create a PowerPoint presentation, you will have considerable flexibility to present your information in a variety of ways for maximum effect. (See pages 217–222 for more information on preparing visual aids.)
- **Rehearse your talk**. The more you rehearse your talk, the smoother your presentation will be when you deliver it to your audience. A couple of timed practice runs will make you aware of any weak points in your talk; they will also let you know if you are running over or under time.

Giving your talk

Dress comfortably
Dressing comfortably means not overdressing but also not dressing down for the occasion. For most student presentations, you should wear your normal clothes. When in doubt, find out what your instructor expects. Scientists are hardly known for their sartorial elegance, so you likely won't need to put a great deal of effort into your appearance to look good. Just don't go too far with informal dress: ripped jeans and an old T-shirt can seem disrespectful to the audience.

Give yourself time at the beginning
If you have equipment to set up or any special preparations to make, try to sort these out before the class begins. If everything is ready to go, you won't get flustered trying to resolve technical problems while your classmates are looking on pityingly.

Begin with an overview
If the audience knows how the talk is structured, they will be able to understand what you are doing as you move from one point to another. Introduce your topic and then give a brief statement of the main areas you will discuss. If your audience is expected to take notes, it is useful to distribute a handout of the PowerPoint presentation.

Project your voice
When you speak, be sure you're loud enough so that everyone in the classroom can hear you. Try looking at the back row of the class and projecting your voice. Also, try to put some feeling into what you say. It is difficult for an audience member to remain attentive to even the most interesting presentation delivered in a monotone. If you've written a dynamic talk, with many short sentences punctuated by a few long sentences, you can develop a rhythm in your speech. However, if you tend to think and speak in long sentences, you will need to put more effort into escaping the trap of monotony.

Don't be apologetic
The worst way to start a talk is by saying, "You'll have to forgive me, I'm really nervous about this" or "I hope this projector is going to work properly." Even if you are nervous, try to create an air of calm confidence. Confidence doesn't mean trying to break the ice with a joke. Although many aspects of the life

sciences lend themselves to humour, opening with a joke or a humorous story can sound apologetic, as it implies that the material needs an entertaining boost. Also, even when jokes don't fail outright they usually come across as strained.

Maintain eye contact with your audience

Look around the room as you speak. When you look at individuals, you involve them in what you are saying. Also, as you scan the faces in front of you, you can monitor them for signs of boredom or incomprehension and can adjust your talk accordingly.

Work with your visual aids

If you have visual aids, take advantage of them; just remember that the visual material should enhance your talk, not deliver it for you. PowerPoint, the most common software used by students, provides a bewildering array of choices. Although there is not enough space in this book to provide a guide on how best to use PowerPoint, you should stick to fonts and formatting styles that help you to tell the story. Use the information on your slide as the base from which you expand your explanation.

Make it as easy as possible for your audience to interpret each visual. As you present each item, give your audience enough time to read through the content. They will find it frustrating to see images or slides flash by before they've had a chance to take in all of the information. When creating slides, keep your text to a minimum—no more than two or three points per slide. Figure 12.1 provides an example of how you can outline the goals of your talk on a simple, easy-to-read slide. In general, try to use plain backgrounds and simple fonts for all visual aids.

If you're presenting figures or graphs, remember to explain the significance of each one. For a graph, first explain the x- and y-axes, and then describe the important points of the graph. Inevitably, your explanations will take more time than you expect. Consider Figure 12.2. While it may be tempting to simply say "ABA and ABA-GE have identical patterns" and move on to the next slide, you need to provide a more complete explanation. In this example, you could start by saying that the slide shows a comparison of the concentrations of two hormones over the course of a year, indicated by months on the x-axis. You should also explain that the hormone concentrations are measured in micrograms per gram dry weight of pine bud material and that each data point is an average of samples gathered from nine different genotypes

Outline of talk

Goal 1: to study multiple hormone profiles of lodgepole pine during cone production, both in natural and experimental conditions

Goal 2: to study efficacy of cone induction treatments

Figure 12.1 Sample slide outlining a talk

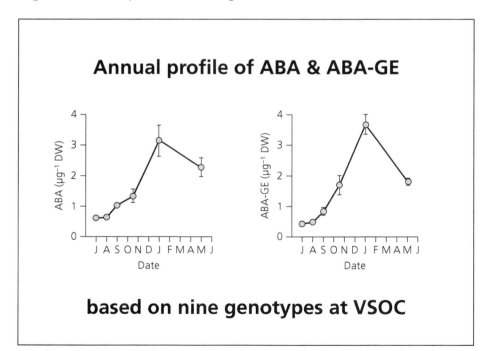

Figure 12.2 Sample slide presenting two graphs

selected from VSOC, which you would have to explain as well (it's the Vernon Seed Orchard Company in Vernon, British Columbia). You would then explain that the two hormones show similar patterns with ABA and ABA-GE concentrations rising to a high in winter and dropping with the spring renewal of plant growth. Once you have described these basics, you can move on to add your interpretations, such as whether this is a typical pattern and whether it's true of all plants or pines, or just this one species of pine.

If you're presenting a diagram, take the audience through it step by step; you may be familiar with the material, but your audience might not be. Figure 12.3 shows a slide that illustrates a complex catabolic pathway. The slide contains a lot of information, but the point of it is that ABA—a plant hormone—is catabolized by one particular pathway, itself just one of five catabolic pathways known for this hormone. In this case, you wouldn't need to explain each pathway, as the audience doesn't need to examine each one to understand your point. Rather, you would only have to say that this slide displays the five known pathways and that the one relevant to your presentation is indicated by the curved arrow. ABA breaks down into 8'OH-ABA, then into PA, and finally into DPA and DPA-GS. As you explain this process, you could add the full

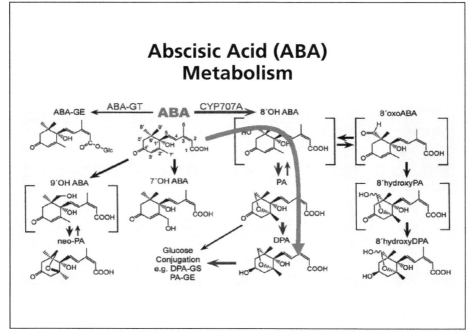

Figure 12.3   Sample slide presenting a diagram

names of the compounds (e.g., 8 hydroxy-ABA) and the major enzyme in the path (CYP707A). While this description is shorter than the description of the graphs in Figure 12.2, it requires a similarly methodical elucidation.

Later in this chapter you will find some tips on how to prepare visual materials to make the strongest impression (see pages 217–222).

Don't go too fast

A good talk is one that is well paced. If you're discussing background information that everyone is familiar with, you can go over it a little faster; if you're describing something complex or less familiar, go slowly. It often helps to explain a complicated point a couple of times in slightly different ways. Don't be afraid to ask your audience if they understand. Almost certainly, someone will speak up if there is a problem.

Pacing is difficult to master, but you should take time to practise, as good pacing can drive your message home. Try to pause before a main point—for example, a conclusion of an experiment—and then spell it out slowly and clearly. Pause again after your explanation to give the audience a moment to reflect on what you have said. Pauses can also make your talk sound more calm and collected. Just as a dynamic combination of notes and rests can enhance a good piece of music, a well-planned combination of discussion and pauses can enhance a good presentation.

Monitor your time allotment

As well as pacing your delivery, you should try to ensure that you aren't going to finish too quickly, or worse, go over your allotted time. If you've rehearsed your talk, you should know roughly how long it will take. Rehearse all of your talk. A common mistake during practice sessions is not practising figure descriptions. These generally take about a minute each and should be done slowly and systematically. Well-explained figures make for much better talks.

Remember to allow extra time for questions that people might ask during your talk. Ideally, you should plan to make your talk a few minutes shorter than the prescribed length so that you have some leeway to answer questions.

End strongly

Don't let your talk fade away at the end. You should finish by summarizing the main points you have made and drawing some conclusions. These conclusions should be available on your visual material so that they can be left there for the discussion. If you can raise some questions in your conclusions, this will set you up for the question period to follow.

Responding to questions

The question period is a time when you can really make a good impression. This is an opportunity for you to demonstrate your thorough understanding of the topic and even to reinforce one or two points that could use further development. If you know your material well, you should have no problem responding to questions, but the manner in which you answer the questions is important, too:

- It's a good idea to repeat a question if you are in a large room where everyone may not have heard it. This should also solidify the question in your mind and give you a few extra moments to consider it before answering.
- If you didn't hear or didn't understand a question, don't be afraid to ask the questioner to repeat or clarify it.
- Keep your answers short and to the point. Rambling answers leave a poor impression.
- Pause before you answer a question. If you know the answer, a pause makes you seem thoughtful. It also makes you appear considerate of the questioner. A pause is a particularly useful device when the question is unexpected, awkward, or even annoying. The pause provides a calm moment in which you can gather yourself.
- If you don't know an answer, say so. It's okay to admit that you don't know everything—as long as you don't do this for every question. It's better to admit that you don't know an answer than to guess or to make up a response that everyone knows is incorrect. If you have no clue, don't admit it openly or be apologetic. It is far better to count silently to three, then ask, "Could you put that question differently?" or, if you need a sure-fire exit strategy say, "I just can't remember at the moment. I'll look that up and get back to you."
- Never answer before someone has completed a question, no matter how long-winded it might be or how certain of the answer you are. It is very important to treat all questions with equal respect.
- If you are asked a question that is aggressive, hostile, or clearly unfair, then simply say, "We can discuss this afterwards."

Dealing with pressure

The nervous speaker
First of all, being nervous is not always a bad thing. It can sharpen the mind remarkably. If you find that you're very nervous just before a talk, take a few deep breaths to lower your heart rate. Then consider what you're going to do about controlling the most obvious signs of nervousness. Don't worry about the signs you can't control, such as skin flushing, nervous tics, and the quaver in your voice—with practice, they will disappear. Rather, focus on what you can control: your hands and your movement. If you're visibly shaky, try placing your hands on a solid surface such as a table or a podium. Keep them there until you need to point to a visual. If you're nervous and your hands aren't steady, avoid using a laser pointer. Laser pointers are small and light; in the hands of a nervous person, the little red dot will jump and careen around the screen, emphasizing your nervousness. If you are nervous enough to develop a dry mouth, try the simple trick known to every actor: pass your tongue over your teeth.

The sensitive speaker
As hard as you try to be informative and entertaining in your talk, not everyone will respond enthusiastically. Indeed, students sometimes roll their eyes, yawn, look bemused, act bored, or even fall asleep during the presentations of their peers. If you're overly sensitive about your audience's reaction to your talk, you might even interpret a neutral gaze on a slumped face, common in every audience, as a lack of interest. Don't read too much into how people look. If you find yourself distracted by such reactions, try to direct your talk to a few audience members who show keen interest. Avoid looking them straight in the eye—this can be distracting as well—but focus on a point between their brows. If you're still worried about keeping your audience's attention, try moving around: walk to the projection screen, point something out on one of your slides, and then walk back to the podium. Such movement will force you to think more about what you're talking about than about who is or isn't listening; it may also draw the attention of some audience members who were starting to drift away from your presentation.

Improving your delivery

It takes a long time to become an effective public speaker. One helpful way to build your skills is to set yourself a goal in each talk. Initially, your goals will probably be simple: finishing on time or finishing on a strong note. Later, you will incorporate more advanced goals: mastering a new technology in the presentation, trying a new organizational style, incorporating feedback on a previous talk, or trying to present from a script. With time, as you adjust your presentation style, you will progress and build confidence. Trying new styles will also help you keep your talks fresh and exciting, both for yourself and for the audience.

Many students wonder whether it is better to be spontaneous or calculated. To take a page out of music, consider the difference between playing jazz and playing classical music. Jazz requires spontaneity and improvisation, while classical music demands calculation and precision. Some musicians prefer jazz over classical music, while others choose classical over jazz. Neither form is inherently *better* than the other, and both can be equally compelling with practice. Similarly, some presenters are able to give expert presentations spontaneously, while others need to plan each step carefully.

As you play around with different presentation styles, consider creating a script as a tool for self-improvement. You may find that you work better with less structure, but creating a detailed script will help you think about how you present and will help you to identify your weaknesses.

Try creating a script
Scripts are easy to put together. Essentially, a script is a written plan of everything that you will say and do in your presentation. The process of scripting may seem counterintuitive, as it would seem to constrain spontaneity. However, once you have created a script, you don't need to read directly from it. Rather, you should memorize it and practise it until it flows naturally.

When it comes time to give your talk, you should bring the script along—if you get lost, you can always pick up your script, get your bearings, and move on with the assurance that you know where you're going. You should set the script in an 18-point font so that you can easily skim the words from a distance and also read it from the podium without squinting. Large type is also easier to read when you're practising. You should highlight prompts so that you can refer to them quickly (see the boldface text in the sample script, below).

The main virtue of a script is that it forces you to think about delivery. Once you have planned *what* you want to say, you can focus on *how* you want to say it. A script provides a structure around which you can place rhetorical devices. These can be elements of timing, repetition (yes, you can repeat yourself to marvellous effect in a lecture), or dynamic clusters of ideas. Winston

Churchill's famous wartime speech "We shall fight them on the beaches" has just such a cluster:

> we shall fight on the beaches,
> we shall fight on the landing grounds,
> we shall fight in the fields and in the streets,
> we shall fight in the hills;
> we shall never surrender. . . .

If you're wondering how this method can be effective in the life sciences, consider the sample script below. This script contains an introduction for a talk on an aspect of tulip biology given by a fictional student named Jane Goderich. Notice that Jane has set the prompts in boldface, for quick reference; she has also underlined words that must be delivered correctly and with emphasis. She repeats words such as *clones* and *cloning* multiple times. Repetition—or thematic reinforcement—is more acceptable in an oral presentation than it is in an essay. Essays and speeches have very different structures. If you want to make a point in a talk, you are welcome to use obvious repetition: "Let me repeat . . ." In fact, repetition of a phrase such as "Let me repeat!" is used in many speeches. If you listen carefully to how people deliver speeches you will become aware that they are able to draw on many different forms of expression.

Start

Show title slide—"Beneath the tulip field"

Do not read the title

My name is Jane Goderich. Today I'm going to explain the biology behind tulip cloning.

Pause

Show slide—"Tulip field in Holland"

In this Dutch field, every plant has the same flower colour, same size, and same leaf shape—same everything. It is a field of identical plants.

Slight pause

Then slowly

It is a field of <u>clones</u>.
Cloning is very common in biology, especially in plant biology.
A big clone is hard to miss—in Utah, there is a single poplar clone that covers 81 ha and weighs in excess of 6,000 t. We think this is the largest organism on earth.

Show slide—"Utah poplar clone"

Another sign that cloning is common can be found in our own language. The rich vocabulary of plant cloning includes dozens of terms: here are just a few.

Show composite slide I—"hyacinth, tiger lily, fern, gladiolus,"

Say rhythmically, while pointing to each one

bulb, bulbil, corm, cormil

Show composite slide II—"strawberry, ginger, potato, fern,"

stolon, runner, tuber, rhizome

Show composite slide III—"moss, hazelnut, poplar, spruce,"

gemma, stool, cutting, ramet

Pause

and my personal favourites

Show composite slide IV—"black cherry, tobacco, cycad"

root sucker, shoot sucker, pup

Slight pause

Tulips clone themselves by <u>offsets</u>.

Show slide—"Tulip with offsets"

Don't rush the next sentences

The development of just such an offset—from initiation to maturity—will be the subject of my talk.
This is the story of a <u>beautiful clone</u>.

    This introduction places tulip cloning within the larger context of plant cloning. It works because it is well organized and structured; it also adds rhythm to what could otherwise be a deadly list of terminology. Notice the small details with which the presenter has described her every step—a good script should contain enough detail that almost anyone could pick it up and make the presentation.
    Since we have a pretty good idea that the body of the talk will be about the biology of offsets, let's skip to the ending.
    You should always try to create a strong ending. The most effective way to end a presentation is with a final sentence that has a sense of closure. You want to let your audience know that the structured presentation is over and open the floor to questions. Usually, you won't need to put up a slide at the end entitled "Questions?" or ask for queries from the audience—questions are inevitable after every talk. Also, don't worry about thanking the audience unless they had to make an effort to attend the talk. Otherwise thanking the audience can sound apologetic and needy, thus weakening your conclusion. The sample ending that follows neither asks for questions nor apologizes. Instead, it gives an example of how audience members might connect the topic to their everyday lives. Of course, there are many ways to end a presentation, and you might have very good reasons for finishing differently.

So, in conclusion . . .

Show slide—"Conclusions": 1. Clones are produced by offsets; 2. The biology of offsets is understood; 3. Only virus-resistant and virus-free clones are selected; 4. Three billion bulbs are exported every year from Holland

Tulips must be bred, then multiplied by offsets, resulting in uniform bulb crops ready to be exported to market.

Pause

The biology of offset growth is relatively simple and lends itself well to large-scale production techniques.

Pause

The strategy in tulip clonal biology is, of necessity, based in combating viruses and other diseases that would otherwise rip through plants of such singular genetic makeup.

Pause

Next time you buy a variety of tulip called Douglas Bader, Toronto, or Pink Diamond, just remember that it is only one of the three billion certified disease-free bulbs that are exported every year from Holland.
Tulips truly do get by with a little help from their friends—humans.

Slide—"Bed of flowering tulips"

As you gain experience working with scripts, you may find that you like having a map of your presentation before you at all times. You may also find that you prefer to work without a set plan. Either way, you will be one step closer to developing your own personal presentation style.

Preparing visual aids for an oral presentation

With the availability of graphic presentation software, as well as laptop computers and video projectors, your ability to use visual aids in a presentation is limited only by your own ingenuity and your instructor's willingness to let you use the technology in class. For instance, you could develop a PowerPoint presentation that includes video clips and sound as well as animated diagrams. Even if you don't have access to PowerPoint or similar software, you can type up the main points of your talk and print them on transparencies along with any pictures and diagrams you have.

Keep it simple

One cardinal rule applies to every aspect of a visual aid—keep it simple. It is much better to put too little material on a slide than too much.

- **Use plain fonts**. Unless you need a fancy one for a specific reason, stick with fonts that are easy to read. Avoid fonts that are too elaborate—they have reduced readability and can become irritating after a few slides. Sans-serif fonts, such as Helvetica and Arial, work well.
- **Use sentence case throughout**. Avoid using ALL CAPS and Title Case (first letters in caps). These styles obscure scientific naming conventions (Latin binomials). From the audience's point-of-view, caps quickly become annoying to read.
- **Use italics for emphasis**. Words set in italics are generally easier to read than words that are underlined.
- **Choose an appropriate font size**. The last thing you want on your slides is text that is too small for the audience to decipher. The 12-point font you use for written assignments will almost certainly be too small when it is projected on a screen. The minimum size you can use will depend to some extent on how far the projector is from the screen, but one rule of thumb suggests 36 points for headings and 24 points for body text. It's always a good idea to test drive your presentation in the room where you'll be presenting your talk so that you can make adjustments if necessary.
- **Be aware of colour sensitivities**. A biological limitation that you might want to consider is that about 8 in 100 men and 1 in 200 women have colour blindness—you might even have a colour sensitivity yourself. In general, avoid using red and green in close proximity to one another.
- **Limit the amount of colour or animation**. Unless you have a good reason for doing so, you should avoid using multicoloured slides or animation effects that are too busy or distracting.
- **Limit the amount of information per slide**. A good rule of thumb is to stick to only four lines of text per slide, not including a heading. White space on a slide is important because it lifts your graph, table, and text—so, keep a lot of it in each slide. Also, if you treat your slides as a script, then you'll be tempted to read directly from them. Instead, make your point briefly on the slide and then expand on the material as you talk. This will make your presentation sound much more natural and professional.

- **Follow the rule of one slide per minute of talk**. Using too many slides can suggest that you are unable to talk without a prop. Take time to establish a rapport with your audience by addressing them in your own words. At the beginning, even if you open your presentation with a slide, try to introduce your topic without pointing to the visual. At the end, after you present your last slide, take a moment to put the focus back on you as a person talking to other people.
- **Complete a computer-based presentation on a single platform**. If possible, avoid transferring your files to multiple computers when using presentation software. Most importantly, if you are switching between a PC and a Mac, or vice versa, make sure it stays identical. Transferring between such different platforms often causes formatting, video, and animation effects to go awry. If you save a PowerPoint presentation in the pptx format, there are significantly fewer problems. If this is not possible, you can also save your presentation as a PDF file.
- **Back up your file frequently**. Students often find themselves engrossed in the formatting and style requirements and forget to save their work at regular intervals. PowerPoint has a habit of seizing up when importing some types of files. Save your work often so that you have previous versions to which you can return in the event of a problem.

Keep it organized

A second fundamental rule of using visual aids is to make sure your material is well organized. If you use a consistent organizational scheme, the audience will become used to it and will be able to follow along more easily.

- **Begin with a title slide**. A title slide sets the tone and orients the audience to your topic. It should contain the title, your name, and the name of the course.
- **Create an outline slide**. An outline slide gives the order of the major sections of your talk so that your audience knows what to expect.
- **Use clear headings**. Headings will help your audience identify the most important point on each slide.
- **Consider section breaks**. If your talk falls naturally into several sections, you could start each section with an introductory slide. Anything that allows the audience to see the structure of your talk is worth including.

- **Keep your overheads in order**. If you are using overhead transparencies, make sure that they are in the correct order—and in the correct orientation—before you start, and be sure to place your transparencies in an ordered pile as you use them. You may have to refer to one later, and you don't want to be shuffling through a disorganized pile in order to find the one you want.
- **Consider passing samples to audience members**. If you're talking about a subject that lends itself to samples, such as snake skins or pine cones, passing them around can be a very effective way to engage the audience in your talk.

The importance of graphics in oral presentations

In the life sciences, talks may rely, sometimes heavily, on photographs, schematics and models. To make a good presentation, visual imagery must not only convey information, but it must also be appealing to the eye. Well-chosen graphical elements can provide power and focus to any talk.

- **Stick to essential graphic elements**. Use only what you really need. Overly complex diagrams, illustrations, graphs, or tables cause viewer fatigue. They are also much harder to talk about because you first must point out what the viewer should concentrate on. Having to do this repeatedly causes boredom and decreases the clarity of your message or interpretation. Make your graphics stand out by leaving ample space around graphics. Remember to keep your text or graphics away from the edges of a PowerPoint slide, as they can get cut off by projectors (see Figure 12.4). The second more important reason is that empty space lifts graphics and improves their viewability.
- **One idea per slide**. In terms of graphics, this usually means sticking to one graphic per slide. After all, you are providing a point of view, usually followed by a reason and an example, often followed by a summary in terms of your point of view. The graphic should contribute to that narrative. If it doesn't, then eliminate it.
- **Always use landscape orientation**. Projection screens and projectors are designed for landscape and not portrait orientation. Don't fight the medium.
- **Plain backgrounds**. Use plain backgrounds, such as white or pastel colours, preferably without a gradient. A plain background will lift graphs and charts. It is possible to use a dark background, such as black, but

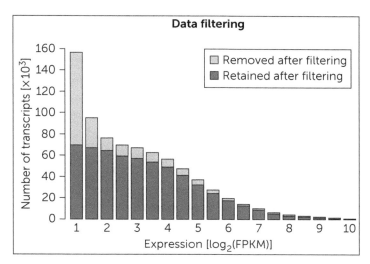

Figure 12.4 The title and the x axis title are both too close to the edge of the slide and run the risk of being cut off by certain types of projectors or even by projection screens. A significant improvement to this slide would to buffer the text and graphic with more white space.

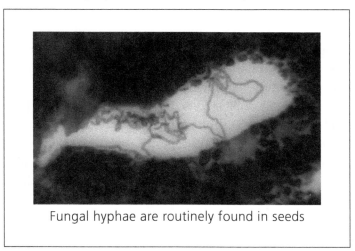

Figure 12.5 This slide is both informative and graphically striking. The fungal hyphae—the threads in the middle—are readily obvious. The statement is brief and informative because it generalizes to all seeds in this presentation. Finally, there is ample white around the graphic, which lifts the illustration.

your choice of text colours is quite limited and some colours must be avoided (e.g. purples, reds, and greens). There is another consideration: namely, the room itself. Dark backgrounds work best in rooms that can

be made fairly dark. However, most presentation rooms are light. Dark text/graphics against a light colour background is a combination that works well everywhere (see Figure 12.5).
- **Templates**. PowerPoint provides many templates, most of which are distracting and even annoying for a science presentation. Either use the plainest template available, or, better still, make up your own plain template.
- **Photographs are either jpeg or gif**. Jpegs tend to be low resolution and are best for full-colour or grey-scale images, whereas gifs are high resolution and are best for text, line art, and solid colours (i.e., a reduced colour palette).
- **Include photographs but make sure to scale them appropriately**. If you add photographs into PowerPoint presentations, try to keep them at around 72 dpi because many large jpegs or gifs will increase the size of a PowerPoint file to a point at which it no longer runs as intended and develops problems that smaller files do not have.
- **Photos and other graphics taken from the web**. You must cite the URL, and if the website requests that you seek permission to use the illustration, then you must get permission.
- **Tables and graphs should always have a title and clear labels.** If you have space, set the text label for the y-axis horizontally—it will be much easier to read. Graphs should be plain, without fill colour, boxes, grid lines, or 3-D effects that could obscure interpretation. If you must show a complex graph or table, describe the overall outcome in brief, then use a blow-up of a part of the graph or table to illustrate your point.
- **Review your presentation out loud.** Reviewing your oral presentation out loud is a good way to discover any graphics that are difficult to explain. If a graphic is giving you the urge to apologize (i.e., "I apologize for this complicated graph/chart/table"), chances are that you should replace it with a simpler graphic or break it into parts that allow easier explanation. Never apologize for a graphic in a talk. Viewers don't like it when speakers make it about themselves. You should be making it about your audience.

Poster Presentations

Posters can describe everything from life cycles to complex models in cell biology. They commonly focus on the results of experimental tests of hypotheses.

Poster presentations are a form of oral presentation—at some point you will have to stand beside your poster and provide a more detailed explanation to interested passers-by. Most students find poster presentations less intimidating than talks. There's nothing nerve-racking about describing your work one-on-one with someone who's genuinely interested. The atmosphere of poster sessions is far less formal as well: they tend to be big, informal social mixers.

Your main goal in creating a poster is to make a visually engaging display. Posters are never put up alone, but in the company of many other posters. Assuming that all participants are presenting equally gripping data or phenomena, how can you make your poster stand out from the crowd? While you are restricted by the limitations of the media—posters are two-dimensional surfaces, most commonly one square metre in size—you should be creative with the visual layout. Focus on using illustrations to replace textual information, providing enough white space for your content to stand out, and choosing an eye-appealing colour scheme. Although this is not meant to be a hard and fast rule, a good general guideline is to create a poster that can be read from two metres away. The three examples in this chapter (pages 227–230) will give you a starting point for creating successful posters.

Using software programs

When you begin to create posters, you will likely use PowerPoint, as it is easy to use, requires few formatting decisions, and is commonly available. As you progress, you should look into more advanced computer software. The best programs for creating posters are those designed for producing graphics that will be printed at publishing industry standards—for example, InDesign, LaTeX, and QuarkXPress. These programs are wonderful because they allow you to fully manipulate and integrate text and graphics in a single document. In addition, you can make lovely posters with many other graphics packages—such as CorelDRAW, OmniGraffle, Inkscape, Illustrator, FreeHand, and PosterGenius to name but a few—some of which are freeware. You will also find *many* poster templates available online—just type "poster template" into your web browser and add the name of the program you are using. Note that the same design principles apply no matter what software you choose.

Creating poster content

Most often, you will create a poster to display the results of an experiment. You should include information that is similar to what you would include in a lab report, but emphasize visual appeal over extensive written descriptions.

- **Abstract**. Normally you do not need to have an abstract on a poster. If you are attending a student conference, a specialized conference, or even an international conference, you will be required to submit an abstract in advance, and it will be published in the conference booklet or on the conference website. By the time you present, many interested attendees will have read your abstract. Include an abstract or summary in your poster only if pre-printed information will not be made available. An abstract on a poster should not exceed 250 words.
- **Introduction**. As in a lab report, you should use the introduction to build up to your hypothesis. Try to keep this section short (under 150 words), and provide just a few references. You can also add an image to draw attention to this section.
- **Materials and Methods**. In posters, you don't need to supply the amount of detail that you would put in a lab report. Flow charts and other illustrations will help you to minimize the text and maximize the clarity of your *Materials and Methods* section. You can always supply the missing details when you describe your poster to an interested viewer.
- **Results**. Lead with your most important result, and then add any necessary explanations of the figures and tables you include in this section. This section will take up the most space on your poster, as it includes the most figures. When presenting your poster to an interested individual, the Results section is when you are traditionally expected to discuss difficulties encountered in your experimental design.
- **Discussion**. This section can be shortened, incorporated in the Results section or Conclusion sections of the poster. The reason Discussion—an otherwise important section in a paper—is not important in a poster is because you are there to discuss its importance with a bystander.
- **Conclusions**. Make sure your conclusions are short and to the point. State whether your hypothesis was supported. Consider your results in terms of just a few published papers and then state the implications your work might have to the field.
- **References**. You can cite—using standardized formats—the references in the poster, but keep the number low, as a *References* section will take up space that you could have put to better use. One way to avoid this section altogether is to provide full reference information in the

text—for example, (Brant & Littlethwaite, 2009. J Poster Graphics 45:11). Toying with a reference forces you to decide whether you really need it. This has the beneficial effect of forcing you to make your text brief, clear, and precise.
- **Acknowledgements.** You will need to thank funding sources, supervisors, and people who provided access to equipment—in short, everyone who made the experiment possible.

Organizing poster content

Keep in mind the following points when organizing your poster:

- **White space.** Measure the area of your poster, and then keep track of how much space you have filled. Good posters are relatively uncluttered, having around 40 per cent white space that is free of illustrations and text.
- **Layout.** Information is presented in columns, dominated by a banner title. Text should be free and not constrained in a box. Because graphs and pictures on the poster represent a "box" or a "block," putting text in boxes framed by lines tends to give an unaesthetic grid-like look.
- **Fonts.** All text should be black and set on a white background—this reads best. For titles and labels, stick with sans-serif fonts, such as Helvetica or Arial. For the main text, use serif fonts, such as Garamond, Palatino, or Times New Roman. The size of the main text should be between 18 and 32 points; subheadings should be between 40 and 60 points; and titles should be 120 points or more. The text should be easily readable from a distance of two metres.
- **Colour scheme.** The photographs or illustrations will dictate some choices, such as using matching or complementary colours to integrate the illustrations into an overall colour scheme. Use a colour wheel to help you choose. Avoid large patches of bright colours as they overwhelm the reader and distract from the message.
- **Illustrations.** Provide labels and captions. Eliminate all excessive information, for example legends on diagrams. Illustrations tend to draw the eye. For that reason, it is best to place them so that the reader is led through the material.

Sample posters

All three of the following sample posters contain worthwhile scientific information, but the second two are far superior in their aesthetic appeal. As you read through each case, try to think about how you would improve upon the poster.

Case 1

This poster was created in PowerPoint (see Figure 12.6). While this program is easy to use, it can be unstable, and professional printers will grimace when told you're giving them a PowerPoint file. Yet the advantage lies in speed: doing a poster in PowerPoint is fast, especially if you use a ready-made template.

The main problem with this poster is that it contains too much written information. It most certainly could not be easily read, even from a metre away! If you were to present this poster at a conference, you would soon learn why too many details can detract from your poster presentation. Little joy comes from having to explain—*ad nauseam*—the contents of an overly detailed poster. Remember: if you give a bad talk to a large audience, you only have to do it once and it's over; if you create a bad poster, you have to present it many, many times over the course of a poster session.

What's good?
- The title, at first blush, appears clear.
- The schematic diagrams are clear and exhibit good use of black and white.
- The text fonts are large enough to read (32 points for text, 44 points for figures, and 60 points for the title) and set in a sans-serif font (Arial).
- There is an acknowledgements section that mentions financial and technical support.

What's bad?
- The title has a colon, creating two clauses, and is done for rhetorical effect. It is better to make your title as concise as possible. A better title would have been "Polarized Defences within Conifer Ovules."
- There are too many logos. Whenever possible, avoid displaying logos on your posters. Logos are institutional branding; they're designed to attract attention, and they're usually graphically superior to your own

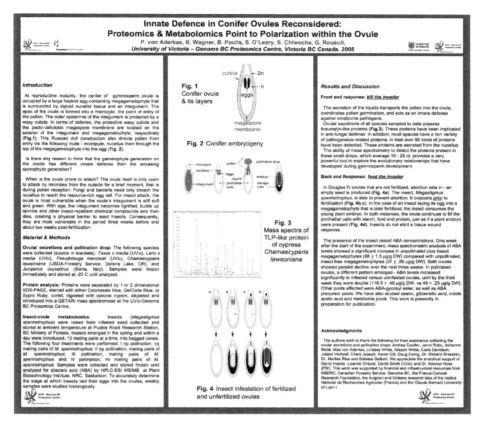

Figure 12.6 Poster for Case 1

illustrations. Therefore, don't let these unnecessary illustrations draw your viewer's attention away from your poster's significant content.

- The layout seems unplanned. It looks as though the poster's creator has merely plugged the information into a template without any further attention to achieving clarity. For example, bulleted lists would have made the text blocks seem less dense. To avoid similar problems, take the time to edit content for conciseness. If you're having trouble seeing the poster objectively, draw up a draft form and have someone else critique its assembly, text, readability, attractiveness, and clarity.
- Some headings are both underlined and set in italics. Underlining makes text difficult to read. When you think you need to underline, use italics instead.

- There is too much text. Each section should have no more than 150 to 200 words.
- There is too little white space. For ease of reading, approximately 30 to 40 per cent of the poster's surface area should be free of text and illustrations.
- It's boring. Imagine yourself as a viewer who is pressed for time. Now imagine standing in front of *this* poster. Would you take the time to examine the content? This is an acid test the poster would not pass.

Case 2

This poster was created in InDesign (see Figure 12.7). Notice how the text flows evenly around the graphics—a layout feature not offered by PowerPoint. This illustrates how you can use advanced software tools to create a better poster.

What's good?
- The poster is easy to read and clear in its goals.
- The title is big and clear—it's an unambiguous question.
- The introduction is illustrated, providing a pair of contrast schematics that explain a critical aspect of the topic without having to reference external literature.
- Rows of dots effectively divide the content into sections.
- Uniformly constructed illustrations of seeds model experimentally derived information.
- The question in the title is linked to the suggested solution at the bottom of the poster. The text in the bottom panel interprets the original question, drives the question forward, and ultimately provides a potential method to resolve the main question.
- The text fonts are perfect: a sans-serif font (Helvetica) for all of the large text (title, subheadings) and a nice serif font (Adobe Garamond) for the main text and the figure titles. Because the main text is in a much bigger font size (48 points) than in the previous poster, the content is more readable.

What's bad?
- It lacks references and acknowledgements.
- There is not enough white space.

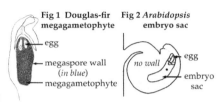

Figure 12.7 Poster for Case 2

Case 3

Another InDesign project, this poster will stand out from its neighbours because it is simple and highly visible (see Figure 12.8).

What's good?
- The white space is sufficient to draw attention to the illustrations and the text.

- The arrows provide easy-to-follow visual paths. These visual tools would also guide the presenter through the presentation steps and help the listener to follow the discussion.
- The illustrations effectively display experimentally derived information, with minimal accompanying text.
- The purpose of the poster is adequately summarized by the title combined with the *Conclusions* section. If you read only the title and the final section, you would get the message of the whole poster in less than 30 seconds.

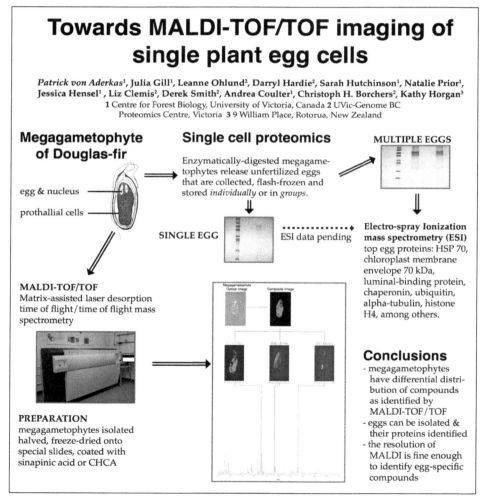

Figure 12.8 Poster for Case 3

- The *Introduction* section is reduced to one picture, *Materials and Methods* is summarized by a few words and a photo, and *Results* is reduced to two pictures and less than 30 words. As you have probably noticed, there isn't even a need for these traditional subheadings.
- The *Conclusions* section consists of an easy-to-skim bulleted list.

What's bad?
- It lacks references and acknowledgements.

Summary

According to some evolutionary psychologists, standing up in front of strangers and talking is a fear that we all share. With experience and practice, you can overcome this fear. Learning how to prepare an organized, articulate, and succinct talk is one important step in this process. Another is learning effective delivery. As in an acting or musical performance, rehearsal and repetition are essential. If you're nervous, you can gain control by taking a few deep breaths and placing your hands on a steady surface. Remember that talks also involve answering questions, and your answers provide a lasting impression. You should always prepare strategies for answering questions of varying difficulty. If you wish to get better at giving oral presentations, you should experiment with different delivery styles (e.g., scripting). If you want to get better at creating posters, you should learn about the principles of graphic design. Poster presentations, compared to oral presentations, are less stressful in the delivery, but you must put much more thought into the visuals.

13 Working in Groups

Objectives
- Learning to work in a group
- Making lab partnerships work
- Keeping a lab book
- Studying in groups
- Working as a group in a lecture class

Group work is important to the development of every young scientist. Group work may involve conducting experiments, researching literature, creating reports, preparing presentations, or putting together many-faceted projects. In the first years of university or college, you will most often participate in group work in lab courses. Although learning scientific facts is often a lonely pursuit, learning to do experiments is more often than not a joint venture.

To work effectively as a group, you must collectively decide how to divide labour and how to support one another. You should appoint one member as the leader who will guide the project to completion. Normally, your lab instructor or professor will provide ample advice on how to set up and maintain an effective group. An effective internal quality control on the work that you have produced is to have your work read by at least one other member of the group. If everyone does this, then the load on the leader is made lighter and the collective effort tends to be better.

Group Dynamics

Groups that work well together have a strong advantage when it comes to completing assignments in full and on time, so it is always in your best interest to get along with your fellow group members.

- An excellent practice is to have a frank discussion at the beginning of the project and again at some point in the term to address the roles, responsibilities, and expectations of your group members.
- To subdivide the work so that it can be most efficiently completed, each individual in a group should highlight their own strengths and weaknesses: the work should be appropriately apportioned.
- The tasks should be assessed at regular intervals by the group so that progress can be measured and subtask completion problems identified before they become a bottleneck to the overall project completion.
- Honest criticism and self-appraisal exercises give students an idea of their relative contribution.
- Comparisons can be quite revealing. They also provide a rational basis for labour redistribution. The result is an efficiently executed project.

Dysfunctional groups are the product of personality conflicts, mutual distrust, incompatible goals, varying abilities, general frustration, and at times bad luck. Sometimes a single person will dominate the discussion—either through arrogance, defiance, or extreme enthusiasm—causing others to become reluctant and withdrawn. In other cases, certain students may be unable to stick to the subject, engaging in unproductive social chatter or recounting loosely related anecdotes for the entertainment of the group. Whatever the problem, you should work to resolve it as soon as possible so you don't waste valuable time.

You can avoid most problems merely by listening to one another—after all, listening is half of the art of conversation. If you find that you are being too forceful in supporting your own opinions, try backing down to give other students a chance to voice their thoughts. You can also build cohesion within your group by openly assessing each other's work. If everyone approaches this task with objectivity and fairness, you should be able to identify the strengths and weaknesses within the group. These assessments will help you to get to know your partners and learn how to work together as a team.

If the problem stems from a general frustration resulting from a seemingly unresolvable problem, try attacking the issue from a different angle to generate new ideas. This process is easier if you appoint a group leader to pose questions and direct the discussion. Try not to give up; most often, banging away at the substance and subtleties of an issue will lead to a breakthrough. If you have enough time and all members of the group are motivated to succeed, you will find that you and your peers can work your way around almost any impasse.

Of course, if you find that you are in a dysfunctional group and you cannot resolve the issues on your own, it is best to tell your instructor. He or she can reshuffle the class into new groups, either by randomly reassigning students or by matching students who have similar personalities and skills.

Lab Partnerships and Groups

As a student, your main objective in participating in lab-based group work is to learn the techniques that will allow you to work and think as a scientist. Your instructor's objective is to guide you and your classmates through this process by directing questions and sharing advice. This is an apprenticeship style of learning—you and your lab partners are novices who share the desire to learn from a master, your instructor. Consequently, everyone involved must work together. You must not only pay close attention to your instructor's directions but also listen carefully to your peers.

Generally, your instructor will assign you a lab partner (or partners) at random. You might think that this practice is unfair, as lab partners can vary in quality, personality, and appeal. Indeed, the most frequent complaint among science students reflects these concerns: "My lab partner is not pulling his or her own weight." While you might fear that your lab partner's work will drag down your own grades, you should always approach a lab partnership with an open mind.

Is randomly grouping students together for lab work an effective team-building method? The answer is unequivocally *yes*. Working with assigned partners—often people who are virtually strangers—dictates that you find new ways to get along for the three or four hours it takes to perform a task, not to mention the hours it takes to prepare a lab report. Working with different groups of people will also help you expand your comfort zone as you learn to trust the abilities of your peers. You will find that the skills you develop in class for getting along with people of diverse abilities and backgrounds will continue to be useful after you graduate.

Achieving success in lab experiments

Before you begin a lab exercise, you should prepare your strategy. First, read up on the methods, visualizing how you will perform each step, and jot down questions. Second, discuss with your lab partners a division of labour and a plan of execution.

- Who will record the process?
- When will you meet to prepare for the experiment?
- Who will carry out each step?
- Who will present the results to your instructor?

Once your experiment is underway, stay calm and focused. This doesn't mean that you can't socialize with your partners; some steps take longer than others, and it is natural to chat. Stay on track by appointing someone to be the human alarm clock whose job is to bring the group's attention back, for example by saying, "Let's focus now. The next step requires some concentration." This same person should also be responsible for identifying when the group is stuck by asking, "What do we do next?"

Before each class, be sure to prepare for questions that your instructor might ask. Preparation will help the questioning process move along smoothly, making the collective learning process more effective. Instructors pose lab-related questions either to individual groups or to the whole class. Some typical examples include the following:

- When you did this, what happened?
- How do you count, measure, or assess . . . ?
- What is your goal in the next part of the experiment?
- What does this term mean?
- What relationship does this result have to what you did earlier?
- What possible mistakes occur with this type of measurement or method?
- Is the sample size big enough?

Assign a member of your group to record the questions and the answers. These questions are not spontaneous; rather, they are designed as teaching tools (i.e., instructors are told to ask these questions), and the answers will be useful to you when you write the lab report.

Polishing group work for good grades

Getting good grades on group work can be challenging. To minimize the possibility of handing in an inadequate assignment, you should formulate a plan that builds in time at the end of the process during which you can edit your work and correct any oversights. Try to leave a day or two to unite the various elements in a clear manner. One member of the group should check through your entire assignment to make certain that your group has addressed and properly integrated all of the requirements. Putting the final polish on these joint efforts is always rewarded. If you're a very organized, detail-oriented person, you should volunteer to do this. If you're not the most organized person of the group, find out who is and ask them to do it. You can still learn from this final step by offering to help that person put everything together.

Separating your work from their work

Often, your instructor will specify exactly what he or she means by "individual effort" in a group project. Every lab course—and sometimes every lab project within a course—will come with its own requirements. If you are uncertain about what you are responsible for, you should immediately contact your instructor. Not doing so may have dire consequences.

However you carry out the experiment, each member of the group must produce individual work showing personal interpretation. Although you are free to discuss a data set, a problem-solving method, or any other aspect of a lab exercise with the other members of your group, you are not free to copy another student's work. To avoid copying, you and your lab partners should each prepare a draft report of the entire assignment. Next, sit down with your group to discuss each person's interpretation. You can still revise your report to reflect important points in this discussion, and you may find that the discussion leads you to a more objective interpretation as you consider the different perspectives of your lab partners. Because you prepared the work individually to begin with, the individual touches will still be apparent after revision.

In labs, it's common for one person to record data while another performs a step in the experiment. Nevertheless, each partner should keep separate notes in his or her own lab book. This way, everyone will have a complete copy of the data when they prepare their individual student reports.

Keeping a lab book

An intact lab book is an important element in recording your experiment. A lab book provides a reliable and undisputable record of what your experiment accomplished. It is a personal record of what you have done, what you have found, and how you have interpreted your findings. It also provides an indication of what each person in a lab group contributed to an experiment.

Your instructors may ask you to keep a separate lab book for each course. In addition, you will need to keep lab books if you work in a lab as a project student, co-op student, honours project student, or summer assistant. If you continue in science after your degree, you will use lab books as a matter of course. The only difference between a lab book that you use in class and one that you use as a paid lab worker is that the former belongs to you, while the latter belongs to the laboratory. Note that in some labs, you might be required to have a supervisor sign off on each page in order to validate the authenticity of the content.

Lab books are taken very seriously among members of the scientific community. Scientists use lab books to keep results, analysis, and protocols that they will use to write future papers or to train future scientists. Lab books contain the details of how scientists turn ideas into experiments that generate data. They are the records of scientific experience. Students in a co-op program who find themselves working as technical assistants in a research lab should be aware that in most industrial and academic labs, failure to keep a lab book is grounds for dismissal. Lab books are almost always kept in the lab and never go home with the researcher. They are irreplaceable and need to be securely stored.

Choose a good lab book

Only use books with a stitched, glued binding. Never use spiral-bound books, binders, or other loose-leaf options in which pages can be easily replaced. Such books call into question the reliability of the information they contain, as undesirable data could be removed and replaced with falsified information and no one would be the wiser. The best lab books lay flat when they are opened. This type of book may be more expensive to buy, but you will come to love the lay-flat feature. Choose a lab book that has numbered pages. For lab work, make sure that the pages in your book are wide enough to allow you to paste in data printouts without folding them. For fieldwork, choose a smaller book that is more convenient to carry around with you.

Keep your records neat and organized

It is important to keep all of the records in your lab book neat and organized. The following tips will help:

- When you first get a lab book, write your name on the spine; put your name, address, and phone number inside the front cover; and write "Table of Contents" at the top of the first few pages, reserving these for a running record of the contents of the book.

- Always write in ink and be sure to choose a pen that won't bleed or run. Do not use pencil in a lab book because it can easily be erased. Erased entries in lab books will cause readers to question your motives for obscuring information. If you are doing fieldwork, you are more likely to have to record data in indelible ink or even in pencil because rain can cause havoc with normal ink.
- Write dates out in full. Abbreviated forms such as 12/11/12 can be very ambiguous. In Canada, the standard form is day, month, year (e.g., 12 November 2012), but some Canadian scientists use the American form of month, day, year (e.g., November 12, 2012).
- Always write legibly, especially when writing numbers. In the event of an error, never erase, white out, or otherwise obscure an entry. Instead, use a ruler to draw a thin line through what you have written. This way you will still be able to read the mistake. Blotting information out completely is as good as never having recorded it: you can't possibly go back and have a look at the errors.
- You should *never* tear pages out of your lab book. Aside from looking untidy, torn pages call into question the validity of the remaining records.

Record everything

As a rule, you should include in your lab book everything related to an experiment. Someone who reads your records must be able to repeat your work exactly. Include your initial thoughts and discussions on the experiment's goals and hypotheses in addition to any background information that you think is useful. Once you start an experiment, record all protocol details, equipment details, names of assistants and partners, sources of chemicals (not just manufacturers, but locations in the lab for your own future reference), software, names of files where you've stored data, and locations of computers that contain those files. In short, record *every* detail that you might need for your analysis. This includes any mistakes you've made—both major and minor. Why do you need to record errors? You must account for and interpret any data that contradict the hypothesis. Are the errors false positives or false negatives? How could they have been avoided? What lessons can be drawn from the failure? The answers to these questions will be very helpful to you when you begin to analyze the results.

Your instructor, TA, or lab supervisor will usually specify the degree of detail you need to include in your records. At a minimum, you should record the details of all calculations, estimations, measurements, and times (e.g., "3:50 p.m.—added tissue to enzyme digestion solution"). When in doubt, think *overkill*. Something that you might think of as a picky detail—such as the brand, model type, and wattage of light bulbs in a growth room—may prove useful later on. Even the type of water you use (tap, distilled, double-distilled, deionized, reverse osmosis–purified, or Milli-Q) is significant. If you include these types of details, your professor will appreciate your efforts, as the best books are those with the most information.

Study Groups

Study groups differ from in-class groups in that they are formed by students, not by instructors. As such, the purpose of a study group is not to learn how to work with your peers but to help each other understand difficult course material. Studying in groups is a highly effective way of learning complex material. Together, like-minded students can often master complex subjects more rapidly than they could on their own.

You will find many advantages in studying with a group:

- First, you will have the opportunity to see subjects from different perspectives. Each member of the group will likely have different strengths and interests—mathematics, biological theory, chemical structure, genetics, etc.—that complement one another, leading the group to a well-rounded understanding of the topic.
- Second, you will find it easier to focus on the task at hand. Most students are better able to avoid procrastination if they have support from peers who share their academic goals.
- Third, you will be forced to verbalize your understanding of different topics. If you can describe what you know in a way that is clear to your fellow group members, you are well on your way to grasping the topic.
- Fourth, your discussions will likely lead to debate, which will prepare you for argumentative questions on your exam.
- Fifth, you will learn, through discussion, how to pronounce the necessary terminology and use it in context. Life science jargon may appear to be another language, but once you can use a complex word such as *Zosterophyllophyta* in a sentence, you own it. As another benefit, learning to speak like a life scientist will help you build confidence in your position within the scientific community.

You will need to put some thought and effort into setting up a study group. Here are some tips for creating a good group:

- Keep the group small (three is a good number); larger groups tend to get distracted easily and chat more frequently.
- Include students who are of the same academic level.
- Pick students who arrive for class on time; punctuality suggests a strong commitment to learning, and these students are less likely to miss or arrive late for a study session.
- Choose students who want to contribute.

Once you choose your members, appoint one person to be in charge of organizing the meeting's location and time. The meeting place should be suitably quiet. Cafés and library study rooms are good, but dorm rooms are usually bad, as they provide many distractions. Each member of the group should turn off his or her cell phone and remain focused on the group's discussion during each meeting. A period of one to one-and-a-half hours is optimal; any longer, and the group's concentration will fade. Decide in advance what topics you will study at each meeting—you will find that simple preplanning will noticeably improve the study group's efficiency.

Group Activities in a Lecture Class

Group activities are becoming more common in lecture classes. Whether a class contains 20 students or 200 students, it is possible for the instructor to create group activities. Usually, groups contain five or fewer members. Large classes used to be difficult to organize into smaller groups, but technological innovations (e.g., clickers and personal electronic devices) have created easier ways for groups of students to interact in a lecture setting.

The purpose of working in a class-based group is to discuss ideas. In some cases, your group will discuss or debate a complex issue. In other cases, your group will need to come up with an answer to a specific problem or question and then present the solution or answer to the class. For the discussion process to work effectively everyone in the group must be included; each member should, in turn, be allowed to venture a thought.

Once you have formed a group, the first step is for each group member to introduce him- or herself. Say your name and give some personal details, such as your major or where you're from. During this phase you should also mention whether or not you have experience with this sort of group activity. Make an effort to include the people with the least experience. The next step is to appoint someone to be the leader. This person will be responsible for speaking on behalf of the group. As a final preparatory step, appoint someone to be the recorder.

To begin the information-gathering process, the leader should set out simple guidelines:

> We need to state our ideas and evaluate them. We'll start by giving each one of you a minute to voice your thoughts. Even if you disagree with someone's opinion, let the speaker finish.

As you present your ideas, try to give an example to illustrate your point. No one should comment on any single opinion or thought until everyone has had a say. During this step, the recorder will take down all the ideas; when the last person has finished, the leader should say:

> Let's go around once more in the same order and see if we're missing anything.

Once the group has generated enough ideas, you can begin the discussion. The leader has to ensure that the group stays on track and within any time limitations; he or she must also ensure that each member of the group feels comfortable and included. The recorder should write down any dissenting opinions, as they will form part of the final response.

The next task is to sort the information. Begin by categorizing each idea as an advantage or a disadvantage (or pro or con), then rank the ideas in order of most to least important. To reach a decision, the leader should pose a series of questions to the group:

> Do we agree that this idea is the most important one? Do we stand behind our ranking? Are there any objections? Is there anything wrong with our proposal?

In the event of a disagreement, the leader should say:

> Shall we go around once more with everyone giving the reason for his or her opinion? We've made real progress, so let's see if we can resolve this issue.

When the group reaches a decision, the leader should thank everyone for their good work.

Peer-Review Exercises

Increasingly, instructors include peer-review exercises in their courses—for example, you may be asked to go over drafts of one another's essays or reports. Being able to recognize the common errors that others make helps you to spot them in your own work. This works best if instructors provide information on how to prioritize criticism. Problems in scholarship, logic, and communication are more important than spelling errors. Equally, changes to your own written work that has been peer-reviewed should begin with the matters of substance and then progress later to correction of errors in style, grammar, and spelling.

Summary

Learning to work in a group has many benefits in the life sciences. Group work will teach you to rely on your peers and to work together to execute tasks that might be too complex or too time-consuming for you to complete on your own. After all, many hands make light work. In the lab, group work will encourage you to incorporate different perspectives into formulating and interpreting an experiment, leading to greater overall objectivity. In order to benefit from a group-learning experience, you must work to get along with all members of your group. Remember to listen to your peers, work with their strengths, and plan each task in advance to avoid unnecessary frustration. The team-based skills you learn in your classes will pave the way to future possibilities. If you work in a lab, you will need to be able to work as part of a team. Basic group skills will also help you to work with people from other disciplines. This ability is essential in the life sciences, where projects are often interdisciplinary, and individuals with diverse expertise are often grouped together to take up new directions.

14 Writing Examinations

Objectives
- Preparing for exams
- Using memory aids
- Asking the right questions
- Identifying special needs
- Writing essay exams
- Writing open-book exams
- Writing take-home exams
- Writing objective (multiple-choice) exams

Most students feel nervous before tests and exams. It's not surprising. Writing an essay exam imposes special pressures. You can't write and rewrite the way you can in a regular essay, you must often write on topics you would otherwise choose to avoid, and you must observe strict time limits. On the surface, objective (multiple-choice) exams may look easier because you don't have to compose the answers, but they force you to be more decisive about your answers than essay exams do and they require very detailed and precise knowledge of the subject. To do your best you need to feel calm—but how? The general guidelines provided in this chapter will help you approach any test or exam with confidence.

Preparing for the Exam

Review regularly

Exam preparation has to begin long before the exam period itself. A weekly review of lecture notes and texts will help you remember important material and relate new information to old. If you don't review regularly, at the end of the semester you'll be faced with relearning rather than remembering. First- and second-year students often struggle with how to develop effective study methods. The answer lies in using different techniques for different types of information.

Set memory triggers

As you review, condense and focus the material by writing down in the margin key words or phrases that will trigger whole sets of details in your mind. The trigger might be a concept word that names or points to an important theory or definition, or it might be a quantitative phrase such as "three causes of the decline in caribou populations" or "five factors in muscle degeneration."

Sometimes you can create an acronym or a learning aid that will trigger an otherwise hard-to-remember set of facts. A learning aid generally takes the form of an *aide-mémoire* (a set of summary notes) or a mnemonic device (a pattern of letters or words that helps you remember a complex series). An acronym is a common type of mnemonic. For example, the acronym PMAT stands for *prophase, metaphase, anaphase, telophase,* and it is doubly useful because it gives both the names of the phases of cell division and the order in which the phases occur. The more items you need to memorize, the more useful mnemonics become. Consider the acronym KPCOFGS, which stands for *kingdom, phylum, class, order, family, genus, species*—the hierarchy of biological classification. An equally effective way to remember this hierarchy is to memorize the nonsense sentence "King Philip came over for grape soda." Another mnemonic sentence that has saved many a student is "Can intelligent Karen solve some foreign mafia operations?" This sentence will help you remember the compounds that make up the Krebs (or tricarboxylic acid) cycle: *citrate, isocitrate, ketoglutarate, succinyl, succinate, fumarate, malate,* and *oxaloacetate*. Who thinks up this stuff? Well, students like you. You can find many more examples online—just type "mnemonic" and "biology" into an online search engine.

Use comparison tables

A lot of exam questions in life sciences courses revolve around comparison. You can master these types of questions by taking the time to sort comparative information into structured tables. Creating comparative tables while you study will also help you to see connections between many themes and concepts that recur throughout the course. Once you understand these connections, you will find exams easier to write.

Use Venn diagrams

Information can be visually structured so that elements—for example, biological entities—are grouped within circles. Those that overlap provide a measure of difference. Multiple-choice exam questions often test a student's ability to discern exactly these differences. Creating your own Venn diagrams is an effective way to visually reinforce comparisons. Venn diagrams are especially useful for picking out transitional groupings of biological organisms in courses that emphasize phylogeny.

Use mind maps

Mind maps are a visual thinking tool used to trace visual and textual information in a diagram. Mind maps have existed in various forms since antiquity. The heart of a mind map is a drawing or image of the topic. It should be in three or more colours. Concepts, ideas, or facts that radiate from the central image should be grouped by colour and arranged individually. In biology, mind maps are very effective for remembering evolutionary relationships, such as the evolution of neural complexity, and for studying centrally important processes, such as photosynthesis. For many students, this link between visual and textual information assists in recall during exams, as the facts are visually connected in a meaningful pattern that is simpler to recall than the isolated details.

Use flash cards and Post-It Notes

Learning jargon in the life sciences is the same as learning words in a foreign language. If you write the definition on one side of a flash card and the word on the other side, then you can easily test your grasp of the different terms. Flash cards are also useful in study groups, as they provide another method that you can use to assess each other's knowledge. If you have an ample sufficiency of Post-Its, you can write the words and definitions on the notes and stick them to places you are likely to read and reread them, such as walls, mirrors, or kitchen bulletin boards.

Ask questions

Think of questions that will get to the heart of the material and force you to examine the relations between various subjects or issues; then think about how you would answer them. The three-C approach discussed on pages 16–18 may help. For example, reviewing the *components* of the subject could mean focusing on the main parts of an issue or on the definitions of major terms or theories. When reviewing *change* in the subject, you might ask yourself what the causes or results of those changes are. To review *context*, you might consider how certain aspects of the subject—issues, theories, actions, results—compare with others in the course. Essentially, the three-C approach forces you to look at the material from different perspectives.

You might get together with friends who are taking the same course and ask each other questions. Just remember that the most useful review questions are not the ones that require you to recall facts but the ones that force you to analyze, integrate, or evaluate information. This sort of preparation works best in a study group (see pages 239–240 for advice on forming an effective study group).

Review online study resources

Copies of past first-year exams and midterms that are available online from Canadian universities and colleges make reasonable practice exams for testing yourself or your friends in a study group. If you cannot find previous exam papers from your institution, consider looking up exams from any Canadian university or college in your own province. First-year biology courses across Canada tend to cover the same material. This is also true for most science subjects. The reason is that tens of thousands of students transfer after their first or second year to another institution. Intercollegiate curriculum committees exist within each province to verify that all institutions are teaching the same biological information, especially in the first year. This ensures that transfer students can continue their degree without having to meet some institutionally idiosyncratic requirement. For upper-level courses, it is best to find past papers from the institution that you're currently attending, as differences between schools become quite marked in the upper years of study.

Talk to your instructor

Instructors and TAs have office hours during which they are prepared to meet with students. You should take advantage of these opportunities. While you may find it more convenient to go online to search for answers, the information you find on the internet is not always reliable. You should also remember that your instructor will create the exam. By speaking directly with your instructor, you will get a clear idea of his or her priorities—knowledge that can help you focus your studying efforts on the most important topics. Don't be shy about going to office hours or worry that you're wasting your instructor's time. Your instructor will likely appreciate your extra efforts in taking the time to discuss course material outside of the classroom.

Identify special needs

Educational institutions make a concerted effort to recognize and accommodate the special needs of students who have learning disabilities such as dyslexia or other perceptual problems or physical disabilities. If you think you require extra assistance, be sure to make your instructor and the appropriate school officials aware of your situation. Professors are used to these situations and should be able to accommodate you. Most institutions also have an office dedicated to assisting students with special needs—all you have to do is contact this office and they will help set everything up. You may be able to complete an exam in a computer lab or under special conditions that will give you the best chance to demonstrate your knowledge.

Allow extra time to get to the exam

Give yourself lots of time to get to the exam. Nothing is more nerve-wracking than thinking you're going to be late because your alarm didn't go off or you got caught in traffic. Remember Murphy's Law: "Whatever can go wrong will." Anticipate any potential difficulties and allow yourself a good margin.

If you are late

In large first- and second-year courses, there are usually rules about lateness. For example, you may be allowed to write only if you arrive within the first half-hour of the exam period; if you arrive any later, you may be banned from the exam. Even if you are late, it is still better to arrive and throw yourself at the mercy of the professor than to not come at all. Most professors will accommodate late students, especially if the reasons for their tardiness are compelling.

Writing an Essay Exam

Read the exam

An exam is not a race. Finishing early, though satisfying, is not usually rewarded with higher grades. Good students tend to fully use the time available. Instead of starting to write immediately, take time at the beginning of the exam to read through each question and create a plan. A few minutes spent on thinking and organizing will bring better results than the same time spent on writing a few more lines.

Apportion your time

Read the instructions carefully to find out how many questions you must answer and to see if you have any choice. Subtract five minutes or so for the initial planning, and then divide the time you have left by the number of questions you have to answer. If possible, allow for a little extra time at the end to reread and edit your work. If the instructions on the exam indicate that not all questions are of equal value, allocate your time accordingly.

Choose your questions

If it's an essay-style exam decide on the questions that you will do and the order in which you will do them. Your answers don't have to be in the same order as the questions. If you think you have lots of time, it's a good idea to place your best answer first, your worst answer in the middle, and your second-best answer at the end, in order to leave the reader on a high note. If you think you will be rushed, though, it's wiser to work from best to worst; that way you will be sure to get all the marks you can on your good answers, and you won't have to cut a good answer short at the end.

Stay calm

If your first reaction on reading the exam is "I can't do any of it!" then force yourself to be calm: take several slow, deep breaths to relax, then decide which question you can answer best. Even if the exam seems impossible at first, you can probably find one question that looks manageable; that's the one to begin with. It will get you rolling and increase your confidence. By the time you have finished your first answer, you will probably find that your mind has calmed enough to tackle other questions.

Read each question carefully

As you turn to each question, read it carefully and underline all the key words. The wording will probably suggest the number of parts your answer should have. Be sure you don't overlook anything—this is a common mistake when people are nervous. Because the verb used in the question is usually a guide for the approach to take in your answer, it's especially important that you interpret the key words in the question correctly. For advice on how to interpret the verbs *outline, trace, explain, discuss, compare,* and *evaluate,* see page 19.

Make notes

Before you begin to organize your answer, jot down key ideas and information related to the topic on extra paper provided or on the spare pages of your answer book. These notes will save you the worry of forgetting something while you are writing. Next, arrange those parts you want to use into a brief plan.

Be direct

Get to the points quickly and use examples to illustrate them. In an exam, as opposed to an essay, it's best to use a direct approach. Don't worry about composing a graceful introduction; simply state the main points that you are going to discuss and then get on with developing them. Remember that your paper will likely be one of many read and marked by someone who has to work quickly; the clearer your answers are, the better they will be received. For each main point, give the kind of specific details that will prove you really know the material. General statements will show you are able to assimilate information, but you need to support these statements with examples to get full marks.

Write legibly

Poor handwriting makes markers cranky. When the person marking your paper has to struggle to decipher your ideas, you may get poorer marks than you deserve. Write on every second or third line of the exam booklet; this will not only make your writing easier to read but also leave you space to make changes and additions if you have time later on. If for some special reason (such as a physical disability) your writing is hard to read, you should be able to make special arrangements to use a computer.

Stick to your time plan

Stay on schedule and don't skip any questions. Try to write something on each topic. Remember that it's easier to score half marks for a question you don't know much about than it is to score full marks for one you could write pages on. If you find yourself running out of time on an answer and still haven't finished, summarize the remaining points and go on to the next question. Leave a large space between questions so that you can go back and add more if you have time.

Reread your answers

No matter how tired or fed up you are, reread your answers at the end if there's time. Check especially for clarity of expression; try to get rid of confusing sentences and improve your transitions so that the logical connections between your ideas are as clear as possible. Revisions that make answers easier to read are always worth the effort.

Writing an Open-Book Exam

If you think that permission to take your books into the exam room is an "open sesame" to success, be forewarned: do not fall into the trap of relying too heavily on your reference materials. You may spend so much time riffling through pages and looking things up that you won't have time to write good answers. The result may be worse than if you had been allowed no books at all.

If you want to do well, use your books only to check information and look up specific, hard-to-remember details for a topic you already know a good deal about. For instance, if your subject is biology, you can look up names and functions; for a biochemistry exam, you can look up reactions; for an exam in health sciences, you can check some references and find the authors' exact definitions of key concepts—if you know where to find them quickly. In other words, use the books to make sure your answers are precise and well-supported but never use them to replace studying and careful exam preparation.

Most instructors will allow you to prepare your book to some degree. This preparation can include slipping in Post-It Notes, adding marginalia (notes pencilled in the margins of the book), and highlighting passages. Some instructors may even allow you to take your own notes into an open-book exam. When this occurs, it pays to take along the comparative tables that you created while studying for the exam.

Writing a Take-Home Exam

The benefit of a take-home exam is that you have time to plan your answers and to consult your textbooks and other sources. The catch is that the amount of time you have to do this is usually less than you would have for a research essay. Don't work yourself into a frenzy trying to respond with a polished essay for each question; instead, aim for well-written exam answers. Keep in mind that you were given this assignment to test your overall command of the course material; your marker is likely to be less concerned with your specialized research than with evidence that you have understood and assimilated the material.

The guidelines for a take-home exam are similar to those for a regular exam; the only difference is that you don't need to keep such a close eye on the clock:

- Keep your introductions short and get to the point quickly.
- Organize your answers in such a way that they are straightforward and clear and the marker can easily see your main ideas.
- Use concrete examples to back up your points.
- Where possible, show the range of your knowledge of course material by referring to a variety of sources rather than constantly using the same ones.
- Try to show that you can analyze and evaluate material—that you can do more than simply repeat information.
- If you are asked to acknowledge the sources of any quotations you use, be sure to jot them down as you go rather than trying to track down sources at the end.

Writing an Objective Exam

In first- and second-year life sciences courses, most exams will consist of objective questions. If you are studying at a large university or college, objective exams may also constitute the bulk of assessment in upper-year courses. Although objective exams sometimes contain true–false questions, they usually feature multiple-choice questions. The main difficulty with these exams is that the questions are designed to confuse the student who is not certain of the correct answers. If you tend to second-guess yourself or if you are the sort of person who readily sees two sides to every question, you may find objective exams particularly hard at first. Fortunately, practice almost always improves performance.

Preparation for objective exams is the same as for other exams. Here, though, it's especially important to pay attention to definitions and unexpected or confusing pieces of information because these are the kinds of details that instructors often use to create questions for objective exams. Although there is no sure recipe for doing well on an objective exam—other than a thorough knowledge of the course material—the following suggestions may help you do better.

Do the easy questions first

Go through the exam at least twice. On the first round, don't waste time on troublesome questions. Because the questions are usually of equal value, it's best to get all the marks you can on the ones you find easy. You can tackle the more difficult questions on the next round. This approach has two advantages: first, you won't be forced, because you have run out of time, to leave out any questions that you could easily have answered correctly; second, when you come back to a difficult question on the second round, you may find that in the meantime you have figured out the answer, or even, in the case of some poorly constructed exams, have seen the answer in the wording of questions further along in the exam.

Find out the marking system

If marks are based solely on the number of right answers, you should pick an answer for every question even if you aren't sure it's the right one. For a true–false question, you have a 50 per cent chance of being right. Even for a multiple-choice question with four possible answers, you have a 25 per cent chance of getting it right, more if you can eliminate one or two of the wrong answers.

On the other hand, if there is a penalty for wrong answers—if marks are deducted for errors—you should guess only when you are fairly sure you are right or when you are able to rule out most of the possibilities. In this case, don't make wild guesses.

Make your guesses educated ones

Before you guess, check that you have read the question correctly. The number one problem on objective exams is misinterpretation of questions. The cure for this problem is to take your time. You could write yourself a reminder on every page of the exam to read carefully; adding this annotation is time well spent and will teach you to approach all questions with equal attention.

Once you know what the question is after, look at the answers. You can approach the choices in one of two ways:

1. decide which answers are truly wrong and see what's left; or
2. if you have studied really well, decide which answers are closest to the correct answer and make your choice from there.

The first approach generally works better—if you can eliminate more than half of the options, you'll have a more manageable number of choices. Once you've narrowed down the options, step away from the question, reread it, and see which one of the remaining answers fits best.

Don't ever make a wild guess unless you're completely stumped by the question itself. Guessing as a regular strategy is *really* bad, because you start guessing at the meaning of questions that, with a little bit of patience, would have become completely obvious. Guessing is the handmaiden of surrender.

If you have to guess, forget about intuition, hunches, and lucky numbers. More importantly, forget about so-called patterns of correct answers—the idea that if there have been two "A" answers in a row, the next one can't possibly be "A" as well, or that if there hasn't been a "true" for a while, "true" must be a good guess. Many question-setters either don't worry about patterns at all or else deliberately confuse pattern-hunters by giving the right answer the same letter or number several times in a row.

Remember that constructing good objective exams is a special skill that not all instructors have mastered. In many cases the questions they pose, though sound enough as questions, do not produce enough realistic alternatives for answers. In such cases the question-setter may resort to some less realistic options, and you can spot them if you pay attention. James F. Shepherd[1] has suggested a number of tips that will increase your chances of making the right guess:

- Start by weeding out all the answers you know are wrong rather than looking for the right one.
- Avoid any terms you don't recognize. Some students are taken in by anything that looks like sophisticated terminology and may assume that such answers must be correct. In fact, these answers are usually wrong; the unfamiliar term may well be a red herring, especially if it is close in sound to the correct one.
- Avoid extremes. Most often the right answer lies in between. For example, suppose that the options are the numbers 800,000; 350,000; 275,000; and 15: the highest and lowest numbers are likely to be wrong.
- Avoid absolutes, especially on questions dealing with people. Few aspects in the life sciences are as certain as is implied by such words as *everyone*, *all*, *no one*, *always*, *invariably*, or *never*. Statements containing these words are usually false.
- Avoid jokes or humorous statements.
- Avoid demeaning or insulting statements. Like jokes, these are usually inserted simply to provide a full complement of options.
- Choose the particular statement over the general (generalizations are usually too sweeping to be true).
- Choose "all of the above" over individual answers. Question-setters know that students with a patchy knowledge of the course material will often fasten on the one fact they know. Only those with a thorough knowledge will recognize that all the answers listed are correct.

Reread the exam

If you have time at the end of the exam, go back and reread the questions. On the other hand, don't start second-guessing yourself and changing a lot of answers at the last minute. Studies have shown that when students make changes they are often wrong. Stick with your original decisions unless you know for certain that you have made a mistake.

Summary

Mastering the exam techniques outlined in this chapter will help you improve your grades in all of your courses. To prepare for exams, you should regularly review course material. In addition to studying on your own, studying with a group can be a very effective way of learning information. Whenever possible, try to find old examination papers to test your knowledge in advance. Comparison tables are particularly useful study tools for life sciences courses—they allow you to see how various features, functions, and concepts relate to one another. If you are writing an essay-type exam, try to answer the easiest question first. You can then tackle the remaining questions in order of increasing difficulty; if you have lots of time, you can put the most difficult answers in the middle and end with another easy question. For an open-book exam, you must study to fully comprehend the material, but you don't need to memorize specific details such as formulae, names, or models. You can prep your book in advance so that it is fingertip ready. When writing a multiple-choice exam, remember to read each question carefully and control the urge to guess. If you are given a take-home exam, you are less constrained by time limitations, but you must still assemble information efficiently to show that you have mastered the material. For each type of exam, you will do best if you prepare well in advance, remain calm and focused during the exam period, and read each question carefully.

Note

1. James F. Shepherd, *College Study Skills*, 6th ed. (Boston, MA: Houghton Mifflin, 2002), and *RSVP: The College Reading, Study, and Vocabulary Program*, 5th ed. (Boston, MA: Houghton Mifflin, 1996).

15 Writing Resumés and Letters of Application

Objectives
- Choosing what to include
- Preparing a standard resumé
- Preparing a functional resumé
- Writing a letter of application
- Applying by email

Whether you are looking for a summer job, applying to graduate school, or seeking permanent employment, eventually you will have to write a resumé and a letter of application. You may even need to write an application letter for some courses or programs, including those with work placement. The person who reads your application will not have time to read reams of material, so you will need to be brief yet precise.

Writing a Resumé

Choosing the best content

Think of a resumé as more than just a summary of facts; think of it as a marketing strategy tailored to specific employers. You will need to supply some basic information, but how you organize it and which details you emphasize are up to you. One good strategy is to put your most important or relevant qualifications first, so that the reader will notice them at first glance. For most students this means leading with educational qualifications, but for others it may mean starting with work experience. Within each section of your resumé, use reverse chronological order so that the most recent item is at the beginning.

Whatever arrangement you choose, your goal is to keep the resumé as concise as possible while including all the specific information that will help you "sell" yourself. A reader will lose interest in a resumé that goes on and on, mixing trivial details with the pertinent ones. On the other hand, experience or skills that may seem irrelevant to you may in fact demonstrate an important attribute or qualification, such as a sense of responsibility or a willingness to work hard. For example, working as a part-time short-order cook may be significant if you state that this was how you paid your way through university or college.

The tone of your resumé should be upbeat, so don't draw attention to any potential weaknesses you may have, such as lack of experience in a particular area. Never list a category and then write "None"—you don't want to suggest that you lack something. Remember to adjust your list of special skills to fit each job you apply for so that the reader will see at a glance that you meet the job requirements. Finally, never claim more for yourself than is true; putting a falsehood into a resumé can be grounds for firing, or even legal action, if it is discovered later.

In Canada you are not required to state your age, place of birth, marital status, race, religion, or sex. Keep in mind, though, that if you are completing an application form from another country, as might occur with a co-op job or graduate school, the legal traditions of that country may well permit their institutions or employers to request information that you would not normally be expected to provide in Canada. In such a case, it is a good idea to provide all the information requested; if you don't, your application may be ignored.

Here is a list of common resumé information, along with some suggestions on how to present it. The examples on pages 259–262 show two different ways of presenting this information depending on the kind of background and experience you have. Most word-processing programs contain a number of templates that will also help you format your resumé.

- **Name.** Typically, you should type your name in capital letters and centre it at the top of the page, although there are many variations on this format.
- **Contact information.** This can include your mailing address with postal code, phone numbers, and email address. If you have a temporary student address, remember to indicate where you can be reached at other times.
- **Career objective** (optional). It's often helpful to let the employer know your career goal, or at least your current aim for employment—for example, "a technician position with opportunity for advancement."
- **Education.** Include any degrees, diplomas, or certificates you have earned, along with the institution that granted them and the date. If it will help your case and if you are short of other qualifications, you should also list courses you have taken that allowed you to develop

skills relevant to the job. In a functional resumé, you may include any training that is additional to your degree under a heading such as "Skills" or "Abilities" (see page 262).
- **Awards or honours.** These may be in a separate section or included with your education details.
- **Work experience.** Give the name and location of your employer, along with your job title and the dates of employment. You don't need to outline all of your duties, but you should list any specialized tasks that you performed. You should focus on listing your accomplishments on the job, using point form and action verbs, for example:

  – Designed and administered a public awareness survey.
  – Supervised a three-member field crew.

  In a functional resumé in the life sciences, you may choose to include work experience under a heading such as "Technical Background" or "Relevant Experience."
- **Research experience or specialized skills.** This section gives you a chance to list information that may give you an advantage in a competitive market, such as experience with certain computer programs or knowledge of a second language. If you have worked as a research assistant, be sure to state the type of work you did and the name of your employer, for example:

  – Assisted Professor Simon Chuong in a laboratory research project, "C4 photosynthesis of *Bienertia*," University of Waterloo, Summer 2018.

- **Other interests** (optional). Depending on the employer and the amount of information you have already included, you may choose to omit this section. Including a few achievements or interests (e.g., travel or athletic accomplishments) will show that you are well rounded; hobbies that require specialized or technical skills (e.g., playing an instrument or building small robots) will suggest coordination, discipline, and attention to detail. Interest in languages shows interest in other cultures, which is an asset in a diverse workplace. If the employer shares some of your interests, they can be a valuable trigger during an interview. Never list interests that involve only passive or minimal participation.
- **Volunteer activities** (optional). Time spent volunteering is a strong measure of community involvement. You should always indicate what

positions you held, what kind of work you did, and the time period that you spent with an organization.
- **References** (optional). Some applicants prefer to wait until they are asked before providing a list of references to employers. If you are granted a job interview, be prepared to provide references on the spot. Have your list neatly printed out and ready to hand to an interviewer; this avoids any delay during which your name could drop down on the list, giving other candidates an advantage. You must contact potential references ahead of time to ensure that they are willing to act in this capacity. Obtain their contact information, including complete email and mailing addresses and phone number. Because reference information is one of the elements that you can control in the hiring process, only provide contacts who are certain to provide you with a good reference. Never use references who will not disclose to you what they would say if contacted.

Preparing a standard resumé

If you're a recent graduate without a great deal of work experience, the standard resumé format will probably show your qualifications in their best light. It includes separate headings for education and work experience and uses a reverse chronological order within those sections. At first, you may need to include all summer and part-time jobs, even if they aren't particularly related to your field. As you gain more experience, you can begin dropping some of the less relevant positions and focusing on those that are significant for the kind of job you're seeking.

ELIZABETH D. LEARNER

Current Address (Until 15 May 2019):
Solin Hall
3510 avenue Lionel-Groulx
Montreal QC H4C 1M7
tel. 514-653-9989
email: liz.d.learner@gmail.com

Permanent Address:
2694 Arbutus Road
Victoria BC V8N 3M2
tel. 250-214-4423

Career objective: An entry-level position with the World Wildlife Fund where my degree in biology would be an asset

Education:
- BSc (Hons) Biology, McGill University (expected June 2019) (Honours thesis: "Predation rates of guppies in Trinidad")

Honours and awards:

- Dean's List, 2016–17, McGill University
- Hugh M. Brock Scholarship ($3,000), 2016, McGill University
- BC Scholarship ($1,500), 2016, British Columbia Scholarship Society

Work experience:

Summer 2018—NSERC USRA research assistant for Professor Brian Falstaff, McGill University, Montreal (Project title: "Nanotechnology of wafer-thin bacterial biscuits")

- Collected bacteria from biofilm surfaces
- Devised, designed, and applied bacteria to nanochips
- Helped design protein calibration technique for method
- Acquired proteomics experience

Summer 2017—research technician, Cellfor Inc., Victoria

- Tissue-cultured Mexican long-leaf pine embryos
- Trained summer students in media preparation

Summer 2016—Server, Sticky Wicket, Victoria

- Waitressed and occasionally tended bar

Specialized skills and experience:

- Aseptic technique in microbiology and biotechnology. Ability to make media, test for contamination, and isolate bacteria as well as plant cells at a proficient level.
- Computer literate. Extensive knowledge of MATLAB; Microsoft Word, Excel, and PowerPoint; Adobe Photoshop and InDesign. Some experience using WordPress.
- Statistical knowledge. Extensive experience with R; moderate experience with SPSS.

Other interests and achievements:

- Trumpeter in Fight Band, McGill, 2017–18
- Member of the intramural women's lacrosse league, McGill, 2017–19
- Musician (bagpipe and tam-tam) at local festivals, 2016–19

Volunteer activities:

- Volunteer for Montreal Children's Hospital, assisting children with disabilities every Saturday since 2016
- Volunteer for SADD (Students Against Drunk Driving) booth at McGill's student orientation, 2018

References (letters available upon request):

Professor Brian Falstaff
(Summer Research Supervisor)
Stewart Biology Building
McGill University
1205 Docteur Penfield
Montreal QC H3A 1B1
Tel. 514-398-0090
brian.falstaff@mcgill.ca

Dr. Diana Tod
(Honours Supervisor)
Redpath Museum
859 Sherbrooke Street West
Montreal QC H3A 2K6
Tel. 514-398-4094
diana.tod@mcgill.ca

Preparing a functional resumé

If your background is less conventional, with a wide range of experience or a change in direction, a functional resumé may be the best format, as it gives greater flexibility to stress transferable skills. Most functional resumés include categories for different areas of expertise (e.g., Research, Administration, Computer Skills). Others may focus on personal attributes such as initiative, teamwork, analytic ability, or communication skills.

GEORGE E.A. CONAPORCUS

Apt. B - 4210 Quinpool Rd.
Halifax NS B3H 3A6
tel. 902-653-9989
email: g.conaporcus@gmail.com

Career objective: Work as a consultant in a marine consulting firm

Profile: Researcher with diverse experience in marine biological systems

Relevant experience:

- Research director for Benthic Directions Inc. (Dartmouth, NS) responsible for day-to-day assignment of work to a team of five research technicians (January 2017–present)
- Technician with Bedford Institute of Oceanography (Dartmouth, NS) in marine bacteriology group (June 2015–December 2017)
- Microbial genomicist with Cranberry Cove Research Inc. (Halifax, NS) responsible for developing chip assays (May 2014–June 2015)

Related skills:
- Currently enrolled part-time in Master of Information Management program, Dalhousie University (started September 2018)
- Certification as a professional diver, Deepwater Diving Inc. (July 2010; certificate available upon request)

Education:

BSc Biochemistry and Molecular Biology, Dalhousie University (2011)

Other achievements and activities:
- Work on marine bacteria in Bedford Basin featured in *The Nature of Things* TV program (March 2015)
- Founder of the St. George's Island Charity (started in 2011) — last year we raised over $55,000 for habitat protection in Halifax Harbour

References:

Available upon request

Writing a Letter of Application

You should *not* use the same letter for all applications; instead, you should craft each one to focus directly on the particular job and company in question and catch the attention of each particular reader. In a sense, both the resumé and the letter of application are intended to open the door to the next stage in the job hunt: the interview. The key is to link your skills to the position, not just to state information. What matters is not what *you want* but what the *employer needs*.

One challenge in writing a letter of application is to tell your reader about yourself and your qualifications without seeming egotistical. Two tips can help:

1. Limit the number of sentences beginning with *I*. Instead, try burying *I* in the middle of some sentences, where it will be less noticeable, for example, "For two months last summer, I worked as a . . . ."
2. Avoid, as much as possible, making unsupported, subjective claims. Instead of saying "I am a highly skilled manager," say something like, "Last summer I managed a $50,000 field study with a crew of seven assistants." Rather than "I have excellent research skills," you might say, "Based on my previous work, Professor Kimiko Sunahara selected me from ten applicants to help with her summer research work."

Here is an example (not to be copied rigidly) of an application letter that tries to connect the applicant's background with the needs of the company.

---

4 March 2019

Jim Newbaggins
Human Resources
BioInc Enterprises
150 Research Road
Manitoba MN R3C 0F4

RE: Job Application for Research Technician

Dear Mr. Newbaggins:

Your advertisement in the *Winnipeg Free Press* for a research technician caught my attention. As a recent graduate of the University of Toronto's Cell and Systems Biology program, I am confident that my qualifications match those you are seeking. The topic of my final thesis was "Proteomics of blood samples from crime scenes." After completing my thesis, I was given a contract to write a small manual to be used in student labs on how to improve handling of time-sensitive biological materials—in particular, protein samples derived from forensic applications. Last year I had the opportunity to speak with your head researcher, Dr. Pantilly, when she came to Toronto and delivered a lecture to my fourth-year Methods in Genomics and Proteomics class. She mentioned that proper handling of samples has become a growing concern as your clientele increases with each year.

Beyond my in-class studies, I have had relevant experience dealing with bio-sampling in a professional lab setting. For the past two summers, I have assisted Dr. Watson of the University of Toronto with her studies of proteomics of mice liver in response to certain pharmaceuticals. My job was to prepare and process liver samples for proteomic analysis. In 2016, I also worked as a co-op student at UGel Inc., a Montreal-based firm specializing in surface films used in the biotechnology industry.

I would appreciate the chance to discuss with you how I could contribute to BioInc Enterprises and will call you next week to see if it is convenient to arrange an interview.

Please note that I have enclosed a resumé.

Sincerely,

Sarah M. Shelston

Sarah M. Shelston

Using Email to Apply for a Position

Many employers welcome email applications. Email has the clear advantage of speed, but it also has some potential pitfalls. The following guidelines will help you avoid them:

- Type your cover letter as an email message, and wherever possible address it to a specific person. Never write "Dear Professor" or "Dear Sir"—you don't want the recipient of your message to mistake it for spam. Indicate whether you are responding to a particular position the organization has advertised rather than applying for any available opportunity. If so, let the "Subject" line at the top of the email immediately tell the reader what you are applying for, such as "Application for Project Manager, file number 360."
- Include your resumé as an attachment. That way, you can be sure the format appears to the reader exactly as you have designed it. If you can, attach the file as a PDF; in this format, your resumé will look the same no matter what type of computer the recipient uses. We all know that the appearance of regular email can sometimes go visually askew at the receiver's end, where fonts, spacing, and alignment can change. An attachment, by contrast, will look the way you created it and can readily be printed and copied in that format. Because you want the cover letter to get immediate attention, leave it as text in the email rather than an attachment.
- Print your message and look at it carefully before pressing the "Send" button. Typographical errors or missing words are often harder to catch on the screen than on the printed page. Another strategy is to send the email to yourself first so that you can see how your message looks and ensure that the attachment opens as it should. Even though email is generally a quick and casual channel of communication, applications are too important to run the risk of error or accident.

> From: liz.d.learner@gmail.com
> Subject: Application for Research Assistant Job in Costa Rica
> Date: 24 January 2019 9:34:57 AM EST (CA)
> To: ayalala@procton.blue.ca Attachment: ELD.pdf
>
> Dear Mr. Ayala:
>
> This message contains my application for the position of research assistant to work on the fauna of bromeliads in Costa Rica, which you recently advertised in the *Canadian Botanical Association Bulletin*. I am nearing completion of my undergraduate Honours Biology degree from McGill University, and I will be available to go into the field as of 15 May 2019. My resumé is attached (see EDL.pdf).
>
> Having twice collected guppies in Trinidad for my Honours thesis under Dr. Tod, I have extensive experience in the tropics. In preparing my thesis, I used various statistical methods to map guppy populations, a skill that is in line with your company's focus on combining ecology with population modelling. I also have experience working in biology labs, both at McGill and at the independent research firm Cellfor Inc. With this experience, I feel that I am well prepared and well qualified to conduct the type of research that you outlined in your advertisement.
>
> Please do not hesitate to contact my references. In addition, I have a valid Canadian passport and a British Columbia driver's licence.
>
> I look forward to hearing from you.
>
> Sincerely,
>
> Elizabeth D. Learner
>
> Current Address:
> Solin Hall
> 3510 Lionel-Groulx
> Montreal QC H4C 1M7
> tel. 514-653-9989
> email: liz.d.learner@gmail.com

Final Words of Advice

When you apply for a job, your application is likely to be one of many. This means that it must pass an initial screening process before it is considered seriously. For that reason, it is absolutely essential that you submit a package that looks professional. Your application package will be judged not just by what you say but also by how you say it. Take the time to double-check for grammar and spelling errors, and make sure that your documents are well formatted. With applications, as with job interviews, first impressions count.

Summary

Resumé writing is a mix of writing to template—standard or functional—and tailoring the content to the employer's needs. Your goal is to display your experience and interests in the best possible light. To this end, you must follow certain style conventions. Always put the most relevant information first. This might be your education, or it might be your experience. Your letter of application, whether it is a printed letter or an email, should tell the employer why you are interested in the job, preparing him or her for your attached resumé. In all cases, you should carefully edit your letter and your resumé before sending your application. When preparing an email submission, be especially careful to use a professional tone. Whereas email, Twitter, and text messages that you send to your friends are spontaneous and informal, communications with potential employers must be formal. Remember that appearance counts—pay attention to details such as content structure, visual layout, and grammar. Because you want to make a good impression, always assemble your application package with care.

Appendix
Weights, Measures, and Notation

BIOLOGY
♂ male
♀ female

STATISTICS
n	sample size
μ	mean of hypothesis
x	mean of sample
σ^2	variance
SE	standard error
SD	standard deviation
CV	coefficient of variation
ANOVA	analysis of variance

UNITS
Quantity
ppm	parts per million
ppb	parts per billion
%	per cent
% solution	mass (g)/100 ml
1 mole (mol)	= 6.022×10^{23} objects
Avogadro's number	= number of objects in a mole (6.022×10^{23})
molar concentration	= molarity, substance concentration, amount concentration
1 molar (1 M)*	= 1 mol/L
1 millimolar (mM)	= 10^{-3} molar
1 micromolar (μM)	= 10^{-6} molar
1 nanomolar (nM)	= 10^{-9} molar
1 picomolar (pM)	= 10^{-12} molar
1 femtomolar (fM)	= 10^{-15} molar
1 attomolar (aM)	= 10^{-18} molar
1 zeptomolar (zM)	= 10^{-21} molar
1 yoctomolar (yM)	= 10^{-24} molar (~1 molecule per 1.6 litres)

Time
1 hour (h)	= 3,600 seconds (s), 60 minutes (min)
1 millisecond (ms)	= 10^{-3} seconds

Length
1 kilometre (km)	= 1,000 metres
1 metre (m)	= 1,000 millimetres
	= 100 centimetres
1 centimetre (cm)	= 10 millimetres (10^{-2} metres)
1 millimetre (mm)	= 1,000 micrometres (10^{-3} metres)
1 nanometre (nm)**	= 10^{-9} metres
1 ångström (Å)	= 10^{-10} metres

Volume (liquid and gas)
1 litre (l or L)	= 1,000 millilitres
1 millilitre (ml or mL)	= 10^{-3} litres
1 cubic centimetre (cm^3)	= 10^{-3} litres
1 microlitre (μl or μL)	= 10^{-6} litres

Mass
1 petagram (Pg)	= 10^{15} grams
1 teragram (Tg)	= 10^{12} grams, or 1 megatonne (Mt)
1 gigagram (Gg)	= 10^9 grams
1 megagram (Mg)	= 10^6 grams, or 1 tonne (t)
1 kilogram (kg)	= 1,000 grams
1 gram (g)	= 1,000 milligrams
1 milligram (mg)	= 1,000 micrograms
1 microgram (μg)	= 10^{-6} grams
1 kilodalton (kDa)	= 1,000 daltons
1 dalton (Da)	= 1.661×10^{-24} grams

Area
1 hectare (ha)	= 10,000 square metres
	= 2.47 acres
1 square metre	= 10,000 square centimetres
	= 1,000,000 square millimetres
1 square centimetre	= 100 square millimetres

*Note: free energy is measured in joules/mole (J/mol).
**Note: wavelengths (λ) are measured in nanometres (e.g., emission λ maximum for blue light is 452–475 nm).

Glossary

abstract. A summary accompanying a formal scientific report or paper, briefly outlining the contents.

abstract language. Language that deals with theoretical, intangible concepts or details: e.g., *complexity, systems approach, truth*. (Compare **concrete language**.)

acronym. A pronounceable word made up of the first letters of the words in a phrase or name: e.g., *FISH* (from *fluorescent in situ hybridization*), *AIDS* (from *acquired immune deficiency syndrome*). A group of initial letters that are pronounced separately is an **initialism**: e.g., *DNA, RNA*.

active voice. See **voice**.

adjectival phrase (or **adjectival clause**). A group of words modifying a noun or pronoun: e.g., *the herring that live in BC waters*.

adjective. A word that modifies or describes a noun or pronoun: e.g., *red, transient, marine*.

adverb. A word that modifies or qualifies a verb, adjective, or adverb, often answering a question such as *how? why? when?* or *where?*: e.g., *slowly, biologically, early, abroad*. (See also **conjunctive adverb**.)

adverbial phrase (or **adverbial clause**). A group of words modifying a verb, adjective, or adverb: e.g., *The cougar ran with great speed*.

agreement. Consistency in tense, number, or person between related parts of a sentence: e.g., between subject and verb, or noun and related pronoun.

ambiguity. Vague or equivocal language; meaning that can be taken two ways.

antecedent (or **referent**). The noun for which a following pronoun stands: e.g., *Blueberries have more anthocyanin compounds when they are mature*.

appositive. A word or phrase that identifies a preceding noun or pronoun: e.g., *Ms. Jones, my TA, is sick*. The second phrase is said to be **in apposition to** the first.

article. See **definite article, indefinite article**.

assertion. A positive statement or claim: e.g., *The data are conclusive*.

auxiliary verb. A verb used to form the tenses, moods, and voices of other verbs: e.g., "am" in *I am recording*. The main auxiliary verbs in English are *be, do, have, can, could, may, might, must, shall, should*, and *will*.

bibliography. (1) A list of works used or referred to in writing an essay or report. (2) A reference book listing works available on a particular subject.

case. Any of the inflected forms of a pronoun (see **inflection**).

    Subjective case. *I, we, you, he, she, it, they*

    Objective case. *me, us, you, him, her, it, them*

    Possessive case. *my/mine, your/yours, our/ours, his, her/hers, its, their/theirs*

circumlocution. A roundabout or circuitous expression, often used in a deliberate attempt to be indirect or evasive: e.g., *kicked the bucket* for "died"; *at this point in time* for "now."

clause. A group of words containing a subject and predicate. An **independent clause** can stand by itself as a complete sentence: e.g., *I measured the scale tips*. A **subordinate** (or **dependent**) **clause** cannot stand by itself but must be connected to another clause: e.g., *Because there were doubts about the first experiment, I measured the scale tips*.

cliché. A phrase or idea that has lost its impact through overuse and betrays a lack of original thought: e.g., *Holy Grail, missing link, paradigm shift*.

collective noun. A noun that is singular in form but refers to a group: e.g., *pack, flock, committee.* It may take either a singular or plural verb, depending on whether it refers to individual members or to the group as a whole.

comma splice. See **run-on sentence.**

complement. A completing word or phrase that usually follows a linking verb to form a **subjective complement**: e.g., (1) *She is my professor*; (2) *That chemical smells terrible.* If the complement is an adjective it is sometimes called a **predicate adjective.** An **objective complement** completes the direct object rather than the subject: e.g., *We found him knowledgeable.*

complex sentence. A sentence containing a dependent clause as well as an independent one: e.g., *I bought the microtome, although it was expensive.*

compound sentence. A sentence containing two or more independent clauses: e.g., *I saw the whale and I photographed it.* A sentence is called **compound–complex** if it contains a dependent clause as well as two independent ones: e.g., *When the fog lifted, I saw the whale and I photographed it.*

conclusion. The part of an essay in which the findings are pulled together or the implications revealed so that the reader has a sense of closure or completion.

concrete language. Specific language that communicates particular details: e.g., *red epidermal cell, three wounded bison.* (Compare **abstract language.**)

conjunction. An uninflected word used to link words, phrases, or clauses. A **coordinating conjunction** (e.g., *and, or, but, for, yet*) links two equal parts of a sentence. A **subordinating conjunction**, placed at the beginning of a subordinate clause, shows the logical dependence of that clause on another: e.g., (1) *Although the dog was small, it was fast*; (2) *While others slept, he studied.* **Correlative conjunctions** are pairs of coordinating conjunctions (see **correlatives**).

conjunctive adverb. A type of adverb that shows the logical relation between the phrase or clause that it modifies and a preceding one: e.g., (1) *I sent the report; it never arrived, however.* (2) *The electricity failed; therefore, the experiment could not begin* .

connotation. The range of ideas or meanings suggested by a certain word in addition to its literal meaning. Apparent synonyms, such as *poor* and *underprivileged*, may have different connotations. (Compare **denotation.**)

context. The text surrounding a particular passage that helps to establish its meaning.

contraction. A word formed by combining and shortening two words: e.g., *isn't* from "is not"; *we're* from "we are."

coordinate construction. A grammatical construction that uses correlatives.

copula verb. See **linking verb.**

correlatives (or **coordinates**). Pairs of correlative conjunctions: e.g., *either/or; neither/nor; not only/but (also).*

dangling modifier. A modifying word or phrase (often including a participle) that is not grammatically connected to any part of the sentence: e.g., *Walking to the birding site, the street was slippery.*

definite article. The word *the*, which precedes a noun and implies that it has already been mentioned or is common knowledge. (Compare **indefinite article.**)

demonstrative pronoun. A pronoun that points out something: e.g., (1) *This is her reason*; (2) *That looks like my forceps.* When used to modify a noun or pronoun, a demonstrative pronoun becomes a **demonstrative adjective**: e.g., *this hat; those people.*

denotation. The literal or dictionary meaning of a word. (Compare **connotation.**)

dependent clause. See **clause.**

diction. The choice of words with regard to their tone, degree of formality, or register. Formal diction is the language of orations and serious essays. The informal diction of everyday speech or conversational writing can, at its extreme, become slang.

direct object. See **object**.

discourse. Talk, either oral or written. **Direct discourse** (or **direct speech**) gives the actual words spoken or written: e.g., *Darwin wrote, "Species are not immutable."* In writing, direct discourse is put in quotation marks. **Indirect discourse** (or **indirect speech**) gives the meaning of the speech rather than the actual words. In writing, indirect discourse is not put in quotation marks: e.g., *He said that species are capable of constant change.*

ellipsis. Three spaced periods indicating an omission from a quoted passage. At the end of a sentence use four periods.

essay. A literary composition on any subject. Some essays are descriptive or narrative, but in an academic setting most are expository (explanatory) or argumentative.

euphemism. A word or phrase used to avoid some other word or phrase that might be considered offensive or blunt: e.g., *pass away* for *die*.

expletive. (1) A word or phrase used to fill out a sentence without adding to the sense: e.g., *To be sure*, *it's not an ideal situation.* (2) A swear word.

exploratory writing. The informal writing done to help generate ideas before formal planning begins.

fused sentence. See **run-on sentence**.

general language. Language that lacks specific details; abstract language.

genes. Gene codes are usually capitalized and either italicized or underlined: e.g., *ADH1*. (Compare **proteins**.) Mutants are usually written in a different case, and may be italicized and/or underlined depending on the organism. Note that conventions can vary greatly between organisms.

gerund. A verbal (part-verb) that functions as a noun and is marked by an *-ing* ending: e.g., *Sequencing can resolve evolutionary relationships.*

grammar. The study of the forms and relations of words and of the rules governing their use in speech and writing.

hypothesis. A supposition or trial proposition made as a starting point for further investigation.

hypothetical instance. A supposed occurrence, often indicated by a clause beginning with *if*.

indefinite article. The word *a* or *an*, which introduces a noun and suggests that it is non-specific. (Compare **definite article**.)

independent clause. See **clause**.

indirect discourse (or **indirect speech**). See **discourse**.

indirect object. See **object**.

infinitive. A type of verbal not connected to any subject: e.g., *to ask*. The **base infinitive** omits the *to*: e.g., *ask*.

inflection. The change in the form of a word to indicate number, person, case, tense, or degree.

initialism. See **acronym**.

intensifier (or **qualifier**). A word that modifies and adds emphasis to another word or phrase: e.g., *very stunted*; *quite large*; *I myself*.

interjection. An abrupt remark or exclamation, usually accompanied by an exclamation mark: e.g., *Oh dear! Alas!*

interrogative sentence. A sentence that asks a question: e.g., *What is the assignment?*

intransitive verb. A verb that does not take a direct object: e.g., *fall, sleep, talk*. (Compare **transitive verb**.)

introduction. A section at the beginning of an essay that tells the reader what is going to be discussed and why.

italics. Slanting type used for emphasis or to indicate the title of a book or journal.

jargon. Technical terms used unnecessarily or in inappropriate places: e.g., *peer-group interaction* for *friendship*.

linking verb (or **copula verb**). A verb such as *be, seem,* or *feel,* used to join subject to complement: e.g., *The apples were ripe.*

literal meaning. The primary, or denotative, meaning of a word.

logical indicator. A word or phrase—usually a conjunction or conjunctive adverb—that shows the logical relation between sentences or clauses: e.g., *since, furthermore, therefore.*

misplaced modifier. A word or group of words that can cause confusion because it is not placed next to the element it should modify: e.g., *I only measured the tips.* [Revised: *I measured only the tips.*]

modifier. A word or group of words that describes or limits another element in the sentence: e.g., *The orchid with the black-striped petals attracted small bees.*

mood. (1) As a grammatical term, the form that shows a verb's function.

   Indicative mood *She is going.*

   Imperative mood *Go!*

   Interrogative mood *Is she going?*

   Subjunctive mood *It is important that she go.*

(2) When applied to literature generally, the atmosphere or tone created by the author.

non-restrictive modifier (or **non-restrictive element**). See **restrictive modifier**.

noun An inflected part of speech marking a person, place, thing, idea, action, or feeling, and usually serving as subject, object, or complement. A **common noun** is a general term: e.g., *dog, chart, chromatograph.* A **proper noun** is a specific name: e.g., *Martin, Sudbury.*

object (1) A noun or pronoun that completes the action of a verb is called a **direct object**: e.g., *She passed the pipettor.* An **indirect object** is the person or thing receiving the direct object: e.g., *She passed Mary* (indirect object) *the pipettor* (direct object). (2) The noun or pronoun in a group of words beginning with a preposition: e.g., *at the lab, about her, for me.*

objective complement. See **complement**.

objectivity. A position or stance taken without personal bias or prejudice. (Compare **subjectivity**.)

outline. With regard to an essay or report, a brief sketch of the main parts; a written plan.

paragraph. A unit of sentences arranged logically to explain or describe an idea, event, or object. The start of a paragraph is sometimes marked by indentation of the first line.

parallel wording. Wording in which a series of items has a similar grammatical form: e.g., *At the beginning of the project, we promised to sample, to analyze, and to store the bacterial DNA.*

paraphrase. Restate in different words.

parentheses. Curved lines enclosing and setting off a passage; not to be confused with square brackets.

parenthetical element. A word or phrase inserted as an explanation or afterthought into a passage that is grammatically complete without it: e.g., *David Suzuki, the second speaker of the day, started his career in genetics before becoming a popularizer of science.*

participle. A verbal (part-verb) that functions as an adjective. Participles can be either **present** (e.g., *speaking to the class*) or **past** (e.g., *spoken before the jury*).

part of speech. Each of the major categories into which words are placed according to their grammatical function. Traditional grammar classifies words based on eight parts of speech: verbs, nouns, pronouns, adjectives, adverbs, prepositions, conjunctions, and interjections.

passive voice. See **voice.**

past participle. See **participle.**

periodic sentence. A sentence in which the normal order is inverted or in which an essential element is suspended until the very end: e.g., *Out of the house, past the grocery store, through the school yard, and down the railway tracks raced the frightened boy.*

person. In grammar, the three classes of personal pronouns referring to the person speaking (**first person**), the person spoken to (**second person**), and the person spoken about (**third person**). With verbs, only the third-person singular has a distinctive inflected form.

personal pronoun. See **pronoun.**

phrase. A unit of words lacking a subject–predicate combination, typically forming part of a clause. The most common kind is the **prepositional phrase**—a unit consisting of a preposition and an object: e.g., *They are waiting at the field station.*

plural. Indicating two or more in number. Nouns, pronouns, and verbs all have plural forms.

possessive case. See **case.**

prefix. An element placed in front of the root form of a word to make a new word: e.g., *pro-, in-, sub-, anti-.* (Compare **suffix.**)

preposition. The introductory word in a unit of words containing an object, thus forming a **prepositional phrase**: e.g., *under the tree; before my time.*

pronoun. A word that stands in for a noun: e.g., *she, this.*

proteins. Proteins codes are capitalized, but neither italicized nor underlined: e.g., ADH1, CYP2E1. (Compare **genes.**)

punctuation. A conventional system of signs (e.g., comma, period, semicolon) used to indicate stops or divisions in a sentence and to make meaning clearer.

reference works. Sources consulted when preparing an essay or report.

referent. See **antecedent.**

reflexive verb. A verb that has an identical subject and object: e.g., *Isabel taught herself to code.*

register. The degree of formality in word choice and sentence structure.

relative clause. A clause introduced by a relative pronoun: e.g., *The man who came to dinner is my honours supervisor.*

relative pronoun. *Who, which, what, that,* or their compounds, used to introduce an adjective or noun clause: e.g., *the research station that David built; whatever you say.*

restrictive modifier (or **restrictive element**). A phrase or clause that identifies or is essential to the meaning of a term: e.g., *The book that I checked out of the library is missing.* It should not be set off by commas. A **non-restrictive modifier** is not needed to identify the term and is usually set off by commas: e.g., *This book, which I checked out of the library, is one of my favourites.*

rhetorical question. A question asked and answered by a writer or speaker to draw attention to a point; no response is expected on the part of the audience: e.g., *How significant are these findings? In my opinion, they are extremely significant, for the following reasons . . .*

run-on sentence. A sentence that goes on beyond the point where it should have stopped. The term covers both the **comma splice** (two sentences incorrectly joined by a comma) and the **fused sentence** (two sentences incorrectly joined without any punctuation).

sentence. A grammatical unit that includes both a subject and a verb. The end of a sentence is marked by a period, question mark, or exclamation mark.

sentence fragment. A group of words lacking either a subject or a verb; an incomplete sentence.

simple sentence. A sentence made up of only one clause: e.g., *The squirrel climbed the tree*.

slang. Colloquial speech considered inappropriate for academic writing; it is often used in a special sense by a particular group: e.g., *dope* for "good" or *diss* for "disrespect."

split infinitive. A construction in which a word is placed between *to* and the base verb: e.g., *to completely finish*. Many still object to this kind of construction, but splitting infinitives is sometimes necessary when the alternatives are awkward or ambiguous.

squinting modifier. A kind of misplaced modifier that could be connected to elements on either side, making meaning ambiguous: e.g., *When he wrote the lab report, finally his TA thanked him*.

standard English. The English currently spoken or written by literate people and widely accepted as the correct and standard form.

subject. In grammar, the noun or noun equivalent with which the verb agrees and about which the rest of the clause is predicated: e.g., *They run programs every day when the room is open*.

subjective complement. See **complement**.

subjectivity. A stance that is based on personal feelings or opinions and is not impartial. (Compare **objectivity**.)

subjunctive. See **mood**.

subordinate clause. See **clause**.

subordinating conjunction. See **conjunction**.

subordination. Making one clause in a sentence dependent on another.

suffix. An element added to the end of a word to form a derivative: e.g., *prepare, preparation; sequence, sequencing*. (Compare **prefix**.)

synonym. A word with the same dictionary meaning as another word: e.g., *begin* and *commence*.

syntax. Sentence construction; the grammatical arrangement of words and phrases.

tense. A set of inflected forms taken by a verb to indicate the time (i.e., past, present, future) of the action.

theme. A recurring or dominant idea.

thesis statement. A one-sentence assertion that gives the central argument of an essay.

topic sentence. The sentence in a paragraph that expresses the main or controlling idea.

transition word. A word that shows the logical relation between sentences or parts of a sentence and thus helps to signal the change from one idea to another: e.g., *therefore, also, however*.

transitive verb. A verb that takes an object: e.g., *hit, bring, cover*. (Compare **intransitive verb**.)

usage. The way in which a word or phrase is normally and correctly used; accepted practice.

verb. That part of a predicate expressing an action, state of being, or condition that tells what a subject is or does. Verbs are inflected to show tense (time). The principal parts of a verb are the three basic forms from which all tenses are made: the base infinitive, the past tense, and the past participle.

verbal. A word that is similar in form to a verb but does not function as one: a participle, a gerund, or an infinitive.

voice. The form of a verb that shows whether the subject acted (**active voice**) or was acted upon (**passive voice**): e.g., *We brought the apparatus* (active). *The apparatus was brought by us* (passive). Only transitive verbs (verbs taking objects) can be passive.

Index

abbreviations, 108, 153; CSE style, 196
Aboriginal, 11
abstract: lab report, 90, 91–2, 104, 105; poster presentation, 224
academic integrity/honesty, 61–4, 188–9
accept/except, 156
accompanied by/with, 156
accuracy: precision and, 98–9
acknowledgements: poster presentations and, 225
acronyms: as study aids, 244
active voice, 118–19; lab report and, 97–8, 100
addresses: punctuation of, 146
adjectives, 135–7; all-purpose, 109; punctuation of, 144–5; use of, 115–16
adverbs, 122–3, 135–7; conjunctive, 153; co-ordinating, 143; use of, 115–16
advice/advise, 156
affect/effect, 157
African Canadian, 11
age: language and, 9, 11
aide-mémoire, 244
all ready/already, 157
all right, 157
all together/altogether, 157
allusion/illusion, 157–8
a lot, 158
alternate/alternative, 158
American Psychological Association style, 190
among/between, 158
amount/number, 158
analysis: essay subject and, 15–19; illustrations and, 171; lab report and, 101; as term, 158
and, 130, 142
anecdote: as essay introduction, 56
anyone/any one, 158
anyways, 158
apostrophe, 140–1
appendices: lab report, 90, 104
application: letter of, 262–3
argumentative essays, 2, 19–20, 22–3
as/because, 159
as/like, 166
assignments: analytical/descriptive, 7
asterisk, 42

as to, 159
as well as, 130
attachments: lab report, 90, 104; resumés and 264
audience, 1, 2–4; oral presentations and, 204, 207, 222; reactions of, 212; *see also* readers
author–date system, 197–200
authors: multiple, 195, 198; organization as, 195, 198; single, 195, 197
axes, graph, 180

backup, 88; presentations and, 219
bad/badly, 159
Baker, Sheridan, and Lawrence B. Gamache, 54–5, 57
bar graphs, 174–5
baseline: bar graphs and, 174
because, 128
because/as, 159
beside/besides, 159
between/among, 158
bibliography, annotated, 200
black (people), 11
blogs, 47; documentation of, 200
books: Chicago style documentation of, 197–8; CSE style documentation of, 195; open-book exams and, 250–1; skimming and, 14
Boolean search operators, 41–2
both. . . and, 122–3, 138
box plots, 177–9
brackets, 141
bring/take, 160
but, 142

calculations: lab report and, 98
can/may, 160
can't hardly, 160
capitals: visual aids and, 218
captions, 184–5; lab report and, 99–100
career objective: resumés and, 257
catalogues, online, 27
categories: essay organization and, 24–5, 52, 54
cause-or-effect analysis: as essay organization, 53

change: three-C approach and, 16, 17–18, 246
chapter: documentation of, 195, 198
checklists: essay, 13; essay editing, 65–6; personal, 66
chemicals: lab report and, 95
Chicago Manual of Style, 190, 197
Chicago style, 197–200; indirect references in, 201
Churchill, Winston, 213–14
circumlocutions, 117
citation maps, 34
citation–name system, 192–3
citation rates, 33–4, 36
citation–sequence system, 191–2, 193
citations: *see* references
cite/sight/site, 160
clarity, 106, 107–15
classification: as essay organization, 52
clauses: independent, 126–7, 153; prepositions and, 134–5; punctuation of, 142–4; reducing, 116–17; relative, 116; restrictive/non-restrictive, 145–6
clichés, 117
coefficient of variation, 98
colon, 60, 141–2
colour: poster presentations and, 225; visual aids and, 218, 220–1
colour sensitivity, 218
comma, 60, 126–7, 142–7
comma splice, 143
common knowledge: quotation and, 64
compare, 19
comparison, 137–8; as essay organization, 53–4
complement/compliment, 160
components: three-C approach and, 16–17, 246
compose/comprise, 160–1
conciseness, 115–17
conclusions: essay, 57–9; figures and, 180; lab report, 90, 103, 105; oral presentation, 210–11, 216–17; poster presentation, 224; summary and, 59
conjunctions, coordinating, 126, 142–3, 147
connotation, 107–8
context: three-C approach and, 16, 18, 246
continual/continuous, 161
contractions, 6–7, 141
contrast: essay organization and, 53–4; writing style and, 122
coordinating conjunctions, 126, 142–3, 147
copyright: illustrations and, 187

correlatives, 122–3, 138
could of, 161
Council of Science Editors (CSE), 190
Council of Science Editors (CSE) style, 191–7; end references in 191, 192, 193–7; indirect references in, 201–2
council/counsel, 161
criterion/criteria, 161–2
critical thinking: research and, 48–9
culture: language and, 9, 10–11

dash, 147–8
data, 162
data analysis: illustrations and, 171
databases: research and, 27–46
dates: abbreviation of, 238; punctuation of, 146, 151; website documentation and, 200
deduce/deduct, 162
defence/defense, 162
definition: as essay organization, 51–2; lab reports and, 90
denotation, 107–8
dependent/dependant, 162
details: concrete, 119–20; lab report and, 97
device/devise, 162
diagrams: lab report and, 96; oral presentations and, 209–10; *see also* illustrations
dictionary, 107; quotation, 56
different from/than, 163
diminish/minimize, 163
disability: language and, 9, 11
discourse: direct, 153; past indirect, 131
discuss, 19
discussion: lecture group and, 240–2
discussion section: lab report, 90, 93, 101–3, 104, 105; poster presentations and, 224
disinterested/uninterested, 163
distortion: illustrations and, 186–7
documentation, 188–202; plagiarism and, 62–4; reasons for, 189–90
DOI (digital object identifier), 194, 199
dollar sign, 42
drafts, first, 50–1
"dry-labbing," 61
due to, 163

each, 129
editing, essay, 65–7
editorials: research and, 43
editors: documentation and, 195, 198

education: resumés and, 256, 257–8
effect/affect, 157
e.g./i.e., 163
either, 129
ellipsis, 148
email: letter of application and, 264–5; research and, 34
emphasis, 120–3; punctuation and, 148, 151
employment: resumés and, 256–66
EndNote, 35
English, 170; plain, 108–9
entomology/etymology, 163
equipment: lab report and, 95–6
errors: illustrations and, 179; lab reports and, 101–2
essays: argumentative, 2, 19–20, 22–3; editing stage of, 65–7; expository, 2, 19, 22–3; format of, 67; organization of, 12, 22, 51–4; persuasive, 2, 19–20, 22–3; planning of, 12–25; research for, 26–49; sample annotated, 67–87; subject and, 15–18; writing of, 50–88
"et al.," 195, 198
evaluate, 19
examinations, 243–55; essay, 248–50; multiple-choice, 243; objective, 243, 252–5; online study resources for, 246; open-book, 250–1; past, 246; preparation for, 243–7; take-home, 251
examples: essays and, 52
except/accept, 156
exceptional/exceptionable, 163–4
exclamation mark, 149
experiments: lab groups and, 234–9; lab reports and, 89, 90–105; research and, 48, 89
explain, 19
explanation: as essay organization, 52
expository essays, 2, 19, 22–3
eye contact: oral presentations and, 207

farther/further, 164
faulty predication, 127–8
fewer/less, 165
figures, 171–87; describing data and, 180; lab report and, 99–100, 102–3; *see also* illustrations
First Nations, 11
flammable/inflammable/non-flammable, 165
flash cards, 245
focus, 164

for, 142
forcefulness, 118–23
formality/informality, 4–7
format: essay, 67; image, 182; lab report, 90–104; references, 190; resumé, 257
frequency distribution: illustration of, 172–3
full circle: as essay conclusion, 58
funnel approach, 54–5

gender: language and, 9–10
generalization: avoiding, 16–17
genomics: language and, 11
gerund, 135
good/well, 164
Google, 29
Google Scholar, 28, 36–7
grades: essays and, 67; group work and, 235–6; objective exams and, 252–3
grammar: common errors in, 125–39; importance of, 139
grammar checkers, 66
graphs, 171–87; axes of, 100, 180; bar, 174–5; describing data and, 180; effective, 172; guidelines for, 100; line, 177; tables and, 185–6; trends and, 180, 186; *see also* illustrations
groups: benefits of, 242; dynamics of, 232–4; dysfunctional, 233–4; lab, 234–9; lecture, 240–2; study, 239–40, 246; working in, 232–42
guess, educated, 253–4
guidelines: graphs, 100; outline, 24–6; writing 8–9

handicapped, 11
handouts: oral presentations and, 205, 206
handwriting, 249–50
hanged/hung, 164
headings: lab reports and, 90; visual aids and, 219
hereditary/heredity, 164
histograms, 175–7; skewed, 176
honesty, academic, 61, 188–9
however, 114–15, 127, 143, 153
human errors, 101–2
humanities style, 197
hyphen, 149
hypothesis, 7–8; formulating, 94–5; lab report and, 93, 101, 102; thesis and, 8, 20

I, 5–6, 262
i.e./e.g., 163

if, 131
illusion/allusion, 157–8
illustrations, 171–87; comparison tables, 244, 251; dangers in, 186–7; describing data and, 180; effective, 172; frequency and category tables, 172–3; guidelines for, 100; lab reports and, 99–101; oral presentations and, 205, 207–8, 220–2; plagiarism and, 64; posters and, 223, 224, 225; simple, 218–19; software for, 99, 185, 186, 217, 223; tables v. graphs, 185–6; working with, 207–10
image formats, 182
in addition to, 130
incite/insight, 165
InDesign, 228
Indian, 11
Indigenous, 11
"individual effort": group work and, 236
infer/imply, 165
inflammable/flammable/non-flammable, 165
inhibit/promote, 168
instructions: essay, 12–13, 19, 53; exam, 248, 249; lab report, 97; details and, 238–9
instructor: role of in exam preparation, 246–7; role of in group work, 234, 235, 236
instrument errors, 101
interlibrary loans, 27
Internet: documentation of, 64, 194, 196, 197, 199–200; evaluation of, 47; research and, 46–7
interpretation: lab report and, 101–2
interruptions: punctuation of, 145, 152
introductions: essay, 51, 54–6; lab report, 90, 92–5, 104, 105; poster presentation, 224; presentation, 3–4
inverse funnel approach, 57–8
irregardless, 165
is when/is where, 128
it, 133
italics, 151–2
It is . . . , 117
its/it's, 141, 165
jargon, 108, 245
job: resumés and 256–66
journals: documentation of, 196, 199; in-house style and, 190; online, 27, 28; skimming and, 14
JSTOR, 27, 28

lab books, 236–9; details and, 238–9; importance of, 237; organization of, 237–8; ownership of, 237
laboratory: research and, 48, 89; *see also* experiments
lab partnerships/groups, 234–9: strategy and, 234–5; *see also* groups
lab reports, 89–105; format of, 90–104; illustrations and, 172; order of sections in, 104–5; writing and, 104–5
language: bias-free, 9–11; clear, 107–10; foreign, 201; *see also* words
leader: group work and, 232, 233; lecture group and, 241–2
lectures: documentation of, 196, 199; group activity in, 240–2
less/fewer, 165
letter of application, 262–3
librarians, 42
library: essay research and, 26–7
lie/lay, 165–6
like/as, 166
line graphs, 177
lists, 139; *see also* series
-ly endings, 150

magazines: documentation of, 196, 199
marking system: objective exams and, 252–3; *see also* grades
materials and methods: lab report, 90, 95–8, 104, 105; poster presentation, 224
may/can, 160
me/myself, 133–4, 166
measures, 267
MEDLINE, 29
memory triggers, 244
method errors, 101
Microsoft Word: images in, 182
might of, 161
mind maps, 245
minimize/diminish, 163
mitigate/militate, 166
MLA Handbook for Writers of Research Papers, 190
mnemonics, 244
modifiers: common errors with, 135–7; dangling, 136–7; restrictive/non-restrictive, 145–6, 168–9; squinting, 135–6

name–year system, 191, 193
Native, 11

Negro, 11
neither, 129
neither . . . nor, 138
new angle: as essay conclusion, 58
New Scientist, 43, 59, 107
newspapers: documentation of, 196, 199; research and, 43
non-flammable/inflammable/flammable, 165
non-restrictive, 168–9
nor/or, 142, 166
not only . . . but also, 122–3, 138
notation, 98, 267
notes: exam writing and, 249; marginal, 251; oral presentations and, 205
notes and bibliography method, 197–200
noun clusters, 116
nouns: collective, 130; pronouns and, 132–3
number/amount, 158
numbers: punctuation of, 142, 151

object: punctuation and, 147
objectivity: lab report and, 89, 98; writing style and, 5–6
off of, 167
only, 136
opinion: essays and, 59
or, 129, 142
oral presentations, 203–22; delivery of, 206; improving delivery of, 212–17; nervousness and, 212; pacing and, 210; preparation for, 204–5; script for, 213–17
organization: citation of, 195, 198; editing and, 65–6; essay, 12, 22, 24–5, 51–4; lab book, 237–8; oral presentation, 219–20; poster, 225
"other interests": resumés and, 258
outline: essay, 22–5; as instruction, 19; oral presentation, 205, 219; writer's block and, 50
overview: presentations and, 206

pairs: common errors with, 137–9
paragraphs: clear, 110–15; coherent, 113–15; focussed, 111–13; idea development and, 111; length of, 115
parallels, 25; common errors with 137–9
parentheses, 42, 152, 191
parenthetical element, 145, 152
passive voice, 5–6, 118–19; gender and, 10; lab report and, 97–8, 100

past indirect discourse, 131
past perfect, 130–1
peer-review, 14, 29; honesty and, 61; Internet and, 43, 47
peer-review exercises, 242
period, 153
periodic sentence, 121
persuasive essay, 2
phenomenon, 167
photographs, 181–3; oral presentations and, 222; resizing, 182–3
phrases: introductory, 146; misused, 156–70
pie charts, 180–1
pixilation, 182–3
plagiarism, 61–4; documentation and, 188, 189; illustrations and, 187; lab report and, 97, 104
planning: essay writing and, 12–25; pros and cons of, 13
populace/populous, 167
possession, 140, 141
poster presentations, 203, 222–31; content of, 223–5; organization of, 225; sample, 226–31; sections of, 224–5
Post-It Notes, 245, 251
PowerPoint, 207, 217, 222, 226
practice/practise, 167
precede/proceed, 167
precision: accuracy and, 98–9
predication, faulty, 128–8
prediction: lab report and, 94, 101
prefixes, 108; hyphens and, 150–1
prepositions: pronouns and, 134–5
prescribe/proscribe, 167
presentations, 203–31; audience for, 3–4; documentation of, 196, 199; *see also* oral presentations; poster presentations
primary sources, 14–15
principle/principal, 167–8
printing: cost of, 185; essays and, 66–7; resumés and, 259, 264
promote/inhibit, 168
pronouns, 113; common errors with, 132–5; first-person, 5–6; gender and, 9–10; objective, 133–4; possessive, 135, 141
Publication Manual of the American Psychological Association, 190
PubMed, 27, 28

punctuation, 140–55; importance of, 155; other punctuation and, 152; quotations and, 60, 141, 142, 147, 148
purpose: lab reports and, 83, 90; statement of, 83

qualifiers, unnecessary, 110
question mark, 42, 153
questions: as essay introduction, 56; essay planning and, 15–18; exam, 248, 249; exam preparation and, 245–6; hypothesis and, 8; indirect, 153; instructor's, 235; journalists', 15–16; multiple-choice, 252, 253; objective exams and, 252–3; oral presentations and, 211; true-false, 252–3
quotation marks, 60, 153–4
quotations: direct, 63; as essay introduction, 55–6; integration of, 59–61; lab report and, 93; length of, 60–1; punctuation of, 60, 141, 142, 147, 148

race: language and, 9, 10–11
rational/rationale, 168
ratios: punctuation of, 142
readers, 1, 2–4; interest of, 55; lab reports and, 90; v. listeners, 3; *see also* audience
reading: out loud, 4, 123
real/really, 168
reason, 128
recorder: lecture group and, 241
reference managers, 202
references, 188–202; essential/nonessential, 193–4; illustrations and, 184, 187; indirect/secondary, 201–2; lab report, 90, 100–1, 104; numbered, 191–2; poster presentations and, 224–5; presentation graphics and, 222; research and, 33–4, 36, 38–41; resumés and, 259
RefWorks, 35
repetition, 3; emphasis and, 123; oral presentations and, 214
research: critical thinking and, 48–9; essays and, 26–49; intersecting subjects and, 37–41; laboratory, 48, 89; into particular scientist, 29–37; resumés and, 258; *see also* experiments
resolution, photo, 182–3
restrictive/non-restrictive modifiers, 145–6, 168–9

results section: lab report, 90, 98–101, 104, 105; poster presentations and, 224
resumés, 256–66; content of, 256–66; email and, 264; falsehoods in, 257; functional, 261–2; legal requirements and, 257; order of, 256; standard, 259–61; tone of, 257
review: exam preparation and, 243–4
review articles, 107
revising, 123; exams and, 250

scatter plots, 179
Scientific American, 43
scientific literature: lab report and, 102
scientific method, 94
scientific notation, 267; lab report and, 98
Scientific Style and Format, 190, 191
scientific terminology, 108
scientists: researching work of, 29–37
search engines, 28, 46–7
search operators, 41–2
seasonable/seasonal, 168
secondary sources, 14–15
semicolon, 127, 153–4
seniors, 11
sentence fragments, 125–6
sentences: grammatical errors in, 125–8; length of, 122; run-on, 126–7, 143; structure of, 121; topic, 111
series, 138–9; complicated, 155; punctuation of, 141–2, 148, 155
Shepherd, James F., 254
should of, 161
sic, 141
sight/site/cite, 160
simple sentence, 121
skills: resumés and, 257, 258
skimming, 14, 37
slang, 5
slides: outline, 219; title, 219; *see also* PowerPoint
so, 142
software: graphing, 99, 185, 186, 217, 223; poster presentations and, 223; referencing, 35, 202, 190; word-processing, 107
sources: documentation of, 188–202; primary, 14–15; secondary, 14–15
special needs: exams and, 247
spell-check, 66
spelling, Canadian, 107

spreadsheet programs, 185
standard deviation, 98
standard error of the mean, 98
study methods, 244–7
style: writing, 106–24
stylistic flourish: as essay conclusion, 58–9
subject: assigned/prescribed, 15; compound, 129–30; essay, 15–18; grammatical, 111–13; identifying, 128–9; intersecting, 37–41; personal, 119; punctuation and, 147
subject–verb agreement, 128–30
subjunctive, 131
subordination, 120–1
suffixes, 108
summary: conclusion and, 59

tables, 172–3, 184; comparison, 244, 251; frequency and category and, 172–3; graphs and, 185–6; *see also* illustrations
take/bring, 160
team-building, 234
tense, 130–2; discussion section and, 102; lab report and, 97, 100
terminology: lab reports and, 90; learning, 245; pronunciation of, 239; scientific, 108
that/which, 168–9
their/there/they're, 169
theme, 19; outline and, 23–4
theory: lab report and, 93
therefore, 127, 143, 153
There is (are)... , 117
thesaurus, 107–8
thesis: developing, 19–22; double-headed, 21; hypothesis and, 8, 20; outline and, 22; precise, 21–2; restricted, 20–1; unified, 21; working, 20, 21–2
thesis statement, 20
"they": as singular, 10
thinking: critical, 48–9; writing and, 1–11
this, 133
three-C approach, 16–18, 245–6
thus, 127, 143
time: exams and, 247, 248, 250, 251; group work and, 235–6; oral presentations and, 206, 210, 219; punctuation of, 142
title page: lab report, 90, 91
titles: illustrations and, 172, 173, 184; punctuation of, 147, 151; verbs and, 130, 132

tone, 4–7
topic: assigned/prescribed, 15, 18–19; broad, 42–6; essay, 15–18; intersecting, 42–6; particular scientist as, 29–37; presentation, 204
topic sentence, 111
tortuous/torturous, 169
trace, 19
translators: documentation and, 198
translucent/transparent, 169
Trimble, John, 51
turbid/turgid, 169

uninterested/disinterested, 163
unique, 169
URL (uniform resource locator), 195, 199
usage: Canadian, 107; common errors in, 125–39

Van Valen, Leigh, 93
variation, coefficient of, 98
Venn diagrams, 245
verbal, 126
verbs: active/passive, 118–19; discussion section and, 102; lab report and, 97, 100; punctuation and, 147; *see also* tense
visual aids, 171–87; simple, 218–19; working with, 207–10; *see also* illustrations
voice: oral presentations and, 206
voice (grammatical): active, 97–8, 100, 118–19; lab report and, 97–8, 100; passive, 5–6, 10, 97–8, 100, 118–19
volunteer activities: resumés and, 258–9

Web of Science, 27–46
Web of Science Core Collection, 29
websites: documentation of, 64, 197, 199–200; *see also* Internet
weights, 267
well/good, 164
whereas, 142
which/that, 168–9
while, 169–70
whiskers, 177–8
who, 170
Wikipedia, 43, 64
wildcards, 42
-wise, 170

words: bias-free, 9–11; clear, 107–10; compound, 150; concrete, 119–10; division of, 149; foreign, 109, 152; Latinate, 108–9; linking, 113–14; misused, 156–70; placement of, 120; plain, 108–9; precise, 109; substitute, 113; transition, 113–14; vague, 21; *see also* language
work experience: resumés and, 256, 258
would of, 161
would/would have, 131
writer's block, 50–1

writing: audience for, 1, 2–4; essay, 50–88; exploratory, 12, 51; guidelines for, 8–9; initial strategies for, 1–8; length of, 4; purpose of, 1–2; structure of, 7–8; style and, 106–24; thinking and, 1–11

yet, 142
your/you're, 170

Zoological Record, 29

The Making Sense Series

Margot Northey with Joan McKibbin
MAKING SENSE
A Student's Guide to Research and Writing
Eighth Edition

Margot Northey, Dianne Draper, and David B. Knight
MAKING SENSE IN GEOGRAPHY AND ENVIRONMENTAL SCIENCES
A Student's Guide to Research and Writing
Sixth Edition

Margot Northey, Lorne Tepperman, and Patrizia Albanese
MAKING SENSE IN THE SOCIAL SCIENCES
A Student's Guide to Research and Writing
Seventh Edition

Margot Northey and Judi Jewinski
MAKING SENSE IN ENGINEERING AND THE TECHNICAL SCIENCES
A Student's Guide to Research and Writing
Fourth Edition

Margot Northey and Patrick von Aderkas
MAKING SENSE IN THE LIFE SCIENCES
A Student's Guide to Research and Writing
Third Edition

Margot Northey, Bradford A. Anderson, and Joel N. Lohr
MAKING SENSE IN RELIGIOUS STUDIES
A Student's Guide to Research and Writing
Third Edition

Margot Northey and Brian Timney
MAKING SENSE IN PSYCHOLOGY
A Student's Guide to Research and Writing
Second Edition

Margot Northey, Kristen Ferguson, and Jon G. Bradley
MAKING SENSE IN EDUCATION
A Student's Guide to Research and Writing
Second Edition